RILEM State-of-the-Art Reports

RILEM STATE-OF-THE-ART REPORTS
Volume 44

RILEM, The International Union of Laboratories and Experts in Construction Materials, Systems and Structures, founded in 1947, is a non-governmental scientific association whose goal is to contribute to progress in the construction sciences, techniques and industries, essentially by means of the communication it fosters between research and practice. RILEM's focus is on construction materials and their use in building and civil engineering structures, covering all phases of the building process from manufacture to use and recycling of materials. More information on RILEM and its previous publications can be found on www.RILEM.net.

The RILEM State-of-the-Art Reports (STAR) are produced by the Technical Committees. They represent one of the most important outputs that RILEM generates – high level scientific and engineering reports that provide cutting edge knowledge in a given field. The work of the TCs is one of RILEM's key functions.

Members of a TC are experts in their field and give their time freely to share their expertise. As a result, the broader scientific community benefits greatly from RILEM's activities.

RILEM's stated objective is to disseminate this information as widely as possible to the scientific community. RILEM therefore considers the STAR reports of its TCs as of highest importance, and encourages their publication whenever possible.

The information in this and similar reports is mostly pre-normative in the sense that it provides the underlying scientific fundamentals on which standards and codes of practice are based. Without such a solid scientific basis, construction practice will be less than efficient or economical.

It is RILEM's hope that this information will be of wide use to the scientific community.

Indexed in SCOPUS, Google Scholar and SpringerLink.

Gabriele Tebaldi · Eshan Dave
Editors

Recycling Asphalt Pavement

State-of-the-Art of the RILEM Technical Committee 264-RAP

Springer

Editors
Gabriele Tebaldi
Department of Engineering
and Architecture
University of Parma
Parma, Italy

Eshan Dave
Department of Civil and Environmental
Engineering
University of New Hampshire
Durham, NH, USA

ISSN 2213-204X ISSN 2213-2031 (electronic)
RILEM State-of-the-Art Reports
ISBN 978-3-031-94230-3 ISBN 978-3-031-94231-0 (eBook)
https://doi.org/10.1007/978-3-031-94231-0

This Springer imprint is published by the registered company Springer Nature Switzerland AG
The registered company address is: Gewerbestrasse 11, 6330 Cham, Switzerland

Members of TC 264-RAP

- Gordon Airey
- Augusto Cannone Falchetto
- Xavier Carbonneau
- Arnaldo Manoel Carneiro
- Eshan V. Dave
- Hervé Di Benedetto
- Vincent Gaudefroy
- Andrea Graziani
- Andrea Grilli
- Gurunath Guduru
- Martin Hugener
- Patricija Kara De Maeijer
- Jan Krol
- Andreas Loizos
- Alexandros Margaritis
- Paul Marsac
- Miomir Miljkovic
- Marko Oreskovic
- Manfred N. Partl
- Emiliano Pasquini
- Katherine Petros
- Laurent Porot
- Simone Raschia
- Francesca Russo
- Margarida Sa da Costa
- Cédric Sauzeat
- Mitsuhiro Shigeishi
- Hassan A. Tabatabaee
- Gabriele Tebaldi
- Wim Van den Bergh
- Kamilla Vasconcelos

- Rosa Veropalumbo
- Nunzio Viscione
- Di Wang
- Michael P. Wistuba
- Huachu Zhai

Collaborators

- Aurelie Lorserie
- Ahmed Abed
- Tony Parry
- Miguel del Sol Sanchez
- Loretta Venturini
- Leda Christiane Lucena
- Luciano Specht
- Goran Mladenovic
- Runhua Zhang
- Gustavo Menegusso Pires
- Gaspare Giancontieri
- Kranthi Kuna
- Yaning Qiao
- Chibuike Ogbo

RILEM Publications

The following list is presents the global offer of RILEM Publications, sorted by series. Each publication is available in printed version and/or in online version.

RILEM Proceedings (PRO)

PRO 1: Durability of High Performance Concrete (ISBN: 2-912143-03-9; e-ISBN: 2-351580-12-5; e-ISBN: 2351580125); *Ed. H. Sommer*

PRO 2: Chloride Penetration into Concrete (ISBN: 2-912143-00-04; e-ISBN: 2912143454); *Eds. L.-O. Nilsson and J.-P. Ollivier*

PRO 3: Evaluation and Strengthening of Existing Masonry Structures (ISBN: 2-912143-02-0; e-ISBN: 2351580141); *Eds. L. Binda and C. Modena*

PRO 4: Concrete: From Material to Structure (ISBN: 2-912143-04-7; e-ISBN: 2351580206); *Eds. J.-P. Bournazel and Y. Malier*

PRO 5: The Role of Admixtures in High Performance Concrete (ISBN: 2-912143-05-5; e-ISBN: 2351580214); *Eds. J. G. Cabrera and R. Rivera-Villarreal*

PRO 6: High Performance Fiber Reinforced Cement Composites—HPFRCC 3 (ISBN: 2-912143-06-3; e-ISBN: 2351580222); *Eds. H. W. Reinhardt and A. E. Naaman*

PRO 7: 1st International RILEM Symposium on Self-Compacting Concrete (ISBN: 2-912143-09-8; e-ISBN: 2912143721); *Eds. Å. Skarendahl and Ö. Petersson*

PRO 8: International RILEM Symposium on Timber Engineering (ISBN: 2-912143-10-1; e-ISBN: 2351580230); *Ed. L. Boström*

PRO 9: 2nd International RILEM Symposium on Adhesion between Polymers and Concrete ISAP '99 (ISBN: 2-912143-11-X; e-ISBN: 2351580249); *Eds. Y. Ohama and M. Puterman*

PRO 10: 3rd International RILEM Symposium on Durability of Building and Construction Sealants (ISBN: 2-912143-13-6; e-ISBN: 2351580257); *Ed. A. T. Wolf*

PRO 11: 4th International RILEM Conference on Reflective Cracking in Pavements (ISBN: 2-912143-14-4; e-ISBN: 2351580265); *Eds. A. O. Abd El Halim, D. A. Taylor and El H. H. Mohamed*

PRO 12: International RILEM Workshop on Historic Mortars: Characteristics and Tests (ISBN: 2-912143-15-2; e-ISBN: 2351580273); *Eds. P. Bartos, C. Groot and J. J. Hughes*

PRO 13: 2nd International RILEM Symposium on Hydration and Setting (ISBN: 2-912143-16-0; e-ISBN: 2351580281); *Ed. A. Nonat*

PRO 14: Integrated Life-Cycle Design of Materials and Structures—ILCDES 2000 (ISBN: 951-758-408-3; e-ISBN: 235158029X); (ISSN: 0356-9403); *Ed. S. Sarja*

PRO 15: Fifth RILEM Symposium on Fibre-Reinforced Concretes (FRC)—BEFIB'2000 (ISBN: 2-912143-18-7; e-ISBN: 291214373X); *Eds. P. Rossi and G. Chanvillard*

PRO 16: Life Prediction and Management of Concrete Structures (ISBN: 2-912143-19-5; e-ISBN: 2351580303); *Ed. D. Naus*

PRO 17: Shrinkage of Concrete—Shrinkage 2000 (ISBN: 2-912143-20-9; e-ISBN: 2351580311); *Eds. V. Baroghel-Bouny and P.-C. Aïtcin*

PRO 18: Measurement and Interpretation of the On-Site Corrosion Rate (ISBN: 2-912143-21-7; e-ISBN: 235158032X); *Eds. C. Andrade, C. Alonso, J. Fullea, J. Polimon and J. Rodriguez*

PRO 19: Testing and Modelling the Chloride Ingress into Concrete (ISBN: 2-912143-22-5; e-ISBN: 2351580338); *Eds. C. Andrade and J. Kropp*

PRO 20: 1st International RILEM Workshop on Microbial Impacts on Building Materials (CD 02) (e-ISBN 978-2-35158-013-4); *Ed. M. Ribas Silva*

PRO 21: International RILEM Symposium on Connections between Steel and Concrete (ISBN: 2-912143-25-X; e-ISBN: 2351580346); *Ed. R. Eligehausen*

PRO 22: International RILEM Symposium on Joints in Timber Structures (ISBN: 2-912143-28-4; e-ISBN: 2351580354); *Eds. S. Aicher and H.-W. Reinhardt*

PRO 23: International RILEM Conference on Early Age Cracking in Cementitious Systems (ISBN: 2-912143-29-2; e-ISBN: 2351580362); *Eds. K. Kovler and A. Bentur*

PRO 24: 2nd International RILEM Workshop on Frost Resistance of Concrete (ISBN: 2-912143-30-6; e-ISBN: 2351580370); *Eds. M. J. Setzer, R. Auberg and H.-J. Keck*

PRO 25: International RILEM Workshop on Frost Damage in Concrete (ISBN: 2-912143-31-4; e-ISBN: 2351580389); *Eds. D. J. Janssen, M. J. Setzer and M. B. Snyder*

PRO 26: International RILEM Workshop on On-Site Control and Evaluation of Masonry Structures (ISBN: 2-912143-34-9; e-ISBN: 2351580141); *Eds. L. Binda and R. C. de Vekey*

PRO 27: International RILEM Symposium on Building Joint Sealants (CD03; e-ISBN: 235158015X); *Ed. A. T. Wolf*

PRO 28: 6th International RILEM Symposium on Performance Testing and Evaluation of Bituminous Materials—PTEBM'03 (ISBN: 2-912143-35-7; e-ISBN: 978-2-912143-77-8); *Ed. M. N. Partl*

PRO 29: 2nd International RILEM Workshop on Life Prediction and Ageing Management of Concrete Structures (ISBN: 2-912143-36-5; e-ISBN: 2912143780); *Ed. D. J. Naus*

PRO 30: 4th International RILEM Workshop on High Performance Fiber Reinforced Cement Composites—HPFRCC 4 (ISBN: 2-912143-37-3; e-ISBN: 2912143799); *Eds. A. E. Naaman and H. W. Reinhardt*

PRO 31: International RILEM Workshop on Test and Design Methods for Steel Fibre Reinforced Concrete: Background and Experiences (ISBN: 2-912143-38-1; e-ISBN: 2351580168); *Eds. B. Schnütgen and L. Vandewalle*

PRO 32: International Conference on Advances in Concrete and Structures 2 vol. (ISBN (set): 2-912143-41-1; e-ISBN: 2351580176); *Eds. Ying-shu Yuan, Surendra P. Shah and Heng-lin Lü*

PRO 33: 3rd International Symposium on Self-Compacting Concrete (ISBN: 2-912143-42-X; e-ISBN: 2912143713); *Eds. Ó. Wallevik and I. Níelsson*

PRO 34: International RILEM Conference on Microbial Impact on Building Materials (ISBN: 2-912143-43-8; e-ISBN: 2351580184); *Ed. M. Ribas Silva*

PRO 35: International RILEM TC 186-ISA on Internal Sulfate Attack and Delayed Ettringite Formation (ISBN: 2-912143-44-6; e-ISBN: 2912143802); *Eds. K. Scrivener and J. Skalny*

PRO 36: International RILEM Symposium on Concrete Science and Engineering—A Tribute to Arnon Bentur (ISBN: 2-912143-46-2; e-ISBN: 2912143586); *Eds. K. Kovler, J. Marchand, S. Mindess and J. Weiss*

PRO 37: 5th International RILEM Conference on Cracking in Pavements—Mitigation, Risk Assessment and Prevention (ISBN: 2-912143-47-0; e-ISBN: 2912143764); *Eds. C. Petit, I. Al-Qadi and A. Millien*

PRO 38: 3rd International RILEM Workshop on Testing and Modelling the Chloride Ingress into Concrete (ISBN: 2-912143-48-9; e-ISBN: 2912143578); *Eds. C. Andrade and J. Kropp*

PRO 39: 6th International RILEM Symposium on Fibre-Reinforced Concretes—BEFIB 2004 (ISBN: 2-912143-51-9; e-ISBN: 2912143748); *Eds. M. Di Prisco, R. Felicetti and G. A. Plizzari*

PRO 40: International RILEM Conference on the Use of Recycled Materials in Buildings and Structures (ISBN: 2-912143-52-7; e-ISBN: 2912143756); *Eds. E. Vázquez, Ch. F. Hendriks and G. M. T. Janssen*

PRO 41: RILEM International Symposium on Environment-Conscious Materials and Systems for Sustainable Development (ISBN: 2-912143-55-1; e-ISBN: 2912143640); *Eds. N. Kashino and Y. Ohama*

PRO 42: SCC'2005—China: 1st International Symposium on Design, Performance and Use of Self-Consolidating Concrete (ISBN: 2-912143-61-6; e-ISBN: 2912143624); *Eds. Zhiwu Yu, Caijun Shi, Kamal Henri Khayat and Youjun Xie*

PRO 43: International RILEM Workshop on Bonded Concrete Overlays (e-ISBN: 2-912143-83-7); *Eds. J. L. Granju and J. Silfwerbrand*

PRO 44: 2nd International RILEM Workshop on Microbial Impacts on Building Materials (CD11) (e-ISBN: 2-912143-84-5); *Ed. M. Ribas Silva*

PRO 45: 2nd International Symposium on Nanotechnology in Construction, Bilbao (ISBN: 2-912143-87-X; e-ISBN: 2912143888); *Eds. Peter J. M. Bartos, Yolanda de Miguel and Antonio Porro*

PRO 46: Concrete Life'06—International RILEM-JCI Seminar on Concrete Durability and Service Life Planning: Curing, Crack Control, Performance in Harsh Environments (ISBN: 2-912143-89-6; e-ISBN: 291214390X); *Ed. K. Kovler*

PRO 47: International RILEM Workshop on Performance Based Evaluation and Indicators for Concrete Durability (ISBN: 978-2-912143-95-2; e-ISBN: 9782912143969); *Eds. V. Baroghel-Bouny, C. Andrade, R. Torrent and K. Scrivener*

PRO 48: 1st International RILEM Symposium on Advances in Concrete through Science and Engineering (e-ISBN: 2-912143-92-6); *Eds. J. Weiss, K. Kovler, J. Marchand, and S. Mindess*

PRO 49: International RILEM Workshop on High Performance Fiber Reinforced Cementitious Composites in Structural Applications (ISBN: 2-912143-93-4; e-ISBN: 2912143942); *Eds. G. Fischer and V. C. Li*

PRO 50: 1st International RILEM Symposium on Textile Reinforced Concrete (ISBN: 2-912143-97-7; e-ISBN: 2351580087); *Eds. Josef Hegger, Wolfgang Brameshuber and Norbert Will*

PRO 51: 2nd International Symposium on Advances in Concrete through Science and Engineering (ISBN: 2-35158-003-6; e-ISBN: 2-35158-002-8); *Eds. J. Marchand, B. Bissonnette, R. Gagné, M. Jolin and F. Paradis*

PRO 52: Volume Changes of Hardening Concrete: Testing and Mitigation (ISBN: 2-35158-004-4; e-ISBN: 2-35158-005-2); *Eds. O. M. Jensen, P. Lura and K. Kovler*

PRO 53: High Performance Fiber Reinforced Cement Composites—HPFRCC5 (ISBN: 978-2-35158-046-2; e-ISBN: 978-2-35158-089-9); *Eds. H. W. Reinhardt and A. E. Naaman*

PRO 54: 5th International RILEM Symposium on Self-Compacting Concrete (ISBN: 978-2-35158-047-9; e-ISBN: 978-2-35158-088-2); *Eds. G. De Schutter and V. Boel*

PRO 55: International RILEM Symposium Photocatalysis, Environment and Construction Materials (ISBN: 978-2-35158-056-1; e-ISBN: 978-2-35158-057-8); *Eds. P. Baglioni and L. Cassar*

PRO 56: International RILEM Workshop on Integral Service Life Modelling of Concrete Structures (ISBN 978-2-35158-058-5; e-ISBN: 978-2-35158-090-5); *Eds. R. M. Ferreira, J. Gulikers and C. Andrade*

PRO 57: RILEM Workshop on Performance of cement-based materials in aggressive aqueous environments (e-ISBN: 978-2-35158-059-2); *Ed. N. De Belie*

PRO 58: International RILEM Symposium on Concrete Modelling—CONMOD'08 (ISBN: 978-2-35158-060-8; e-ISBN: 978-2-35158-076-9); *Eds. E. Schlangen and G. De Schutter*

PRO 59: International RILEM Conference on On Site Assessment of Concrete, Masonry and Timber Structures—SACoMaTiS 2008 (ISBN set: 978-2-35158-061-5; e-ISBN: 978-2-35158-075-2); *Eds. L. Binda, M. di Prisco and R. Felicetti*

PRO 60: Seventh RILEM International Symposium on Fibre Reinforced Concrete: Design and Applications—BEFIB 2008 (ISBN: 978-2-35158-064-6; e-ISBN: 978-2-35158-086-8); *Ed. R. Gettu*

PRO 61: 1st International Conference on Microstructure Related Durability of Cementitious Composites 2 vol., (ISBN: 978-2-35158-065-3; e-ISBN: 978-2-35158-084-4); *Eds. W. Sun, K. van Breugel, C. Miao, G. Ye and H. Chen*

PRO 62: NSF/ RILEM Workshop: In-situ Evaluation of Historic Wood and Masonry Structures (e-ISBN: 978-2-35158-068-4); *Eds. B. Kasal, R. Anthony and M. Drdácký*

PRO 63: Concrete in Aggressive Aqueous Environments: Performance, Testing and Modelling, 2 vol., (ISBN: 978-2-35158-071-4; e-ISBN: 978-2-35158-082-0); *Eds. M. G. Alexander and A. Bertron*

PRO 64: Long Term Performance of Cementitious Barriers and Reinforced Concrete in Nuclear Power Plants and Waste Management—NUCPERF 2009 (ISBN: 978-2-35158-072-1; e-ISBN: 978-2-35158-087-5); *Eds. V. L'Hostis, R. Gens and C. Gallé*

PRO 65: Design Performance and Use of Self-consolidating Concrete—SCC'2009 (ISBN: 978-2-35158-073-8; e-ISBN: 978-2-35158-093-6); *Eds. C. Shi, Z. Yu, K. H. Khayat and P. Yan*

PRO 66: 2nd International RILEM Workshop on Concrete Durability and Service Life Planning—ConcreteLife'09 (ISBN: 978-2-35158-074-5; ISBN: 978-2-35158-074-5); *Ed. K. Kovler*

PRO 67: Repairs Mortars for Historic Masonry (e-ISBN: 978-2-35158-083-7); *Ed. C. Groot*

PRO 68: Proceedings of the 3rd International RILEM Symposium on 'Rheology of Cement Suspensions such as Fresh Concrete (ISBN 978-2-35158-091-2; e-ISBN: 978-2-35158-092-9); *Eds. O. H. Wallevik, S. Kubens and S. Oesterheld*

PRO 69: 3rd International PhD Student Workshop on 'Modelling the Durability of Reinforced Concrete (ISBN: 978-2-35158-095-0); *Eds. R. M. Ferreira, J. Gulikers and C. Andrade*

PRO 70: 2nd International Conference on 'Service Life Design for Infrastructure' (ISBN set: 978-2-35158-096-7, e-ISBN: 978-2-35158-097-4); *Eds. K. van Breugel, G. Ye and Y. Yuan*

PRO 71: Advances in Civil Engineering Materials—The 50-year Teaching Anniversary of Prof. Sun Wei' (ISBN: 978-2-35158-098-1; e-ISBN: 978-2-35158-099-8); *Eds. C. Miao, G. Ye and H. Chen*

PRO 72: First International Conference on 'Advances in Chemically-Activated Materials—CAM'2010' (2010), 264 pp., ISBN: 978-2-35158-101-8; e-ISBN: 978-2-35158-115-5; *Eds. Caijun Shi and Xiaodong Shen*

PRO 73: 2nd International Conference on 'Waste Engineering and Management—ICWEM 2010' (2010), 894 pp., ISBN: 978-2-35158-102-5; e-ISBN: 978-2-35158-103-2, *Eds. J. Zh. Xiao, Y. Zhang, M. S. Cheung and R. Chu*

PRO 74: International RILEM Conference on 'Use of Superabsorbent Polymers and Other New Additives in Concrete' (2010) 374 pp., ISBN: 978-2-35158-104-9; e-ISBN: 978-2-35158-105-6; *Eds. O.M. Jensen, M.T. Hasholt, and S. Laustsen*

PRO 75: International Conference on 'Material Science—2nd ICTRC—Textile Reinforced Concrete—Theme 1' (2010) 436 pp., ISBN: 978-2-35158-106-3; e-ISBN: 978-2-35158-107-0; *Ed. W. Brameshuber*

PRO 76: International Conference on 'Material Science—HetMat—Modelling of Heterogeneous Materials—Theme 2' (2010) 255 pp., ISBN: 978-2-35158-108-7; e-ISBN: 978-2-35158-109-4; *Ed. W. Brameshuber*

PRO 77: International Conference on 'Material Science—AdIPoC—Additions Improving Properties of Concrete—Theme 3' (2010) 459 pp., ISBN: 978-2-35158-110-0; e-ISBN: 978-2-35158-111-7; *Ed. W. Brameshuber*

PRO 78: 2nd Historic Mortars Conference and RILEM TC 203-RHM Final Workshop—HMC2010 (2010) 1416 pp., e-ISBN: 978-2-35158-112-4; *Eds. J. Válek, C. Groot and J. J. Hughes*

PRO 79: International RILEM Conference on Advances in Construction Materials Through Science and Engineering (2011) 213 pp., ISBN: 978-2-35158-116-2, e-ISBN: 978-2-35158-117-9; *Eds. Christopher Leung and K.T. Wan*

PRO 80: 2nd International RILEM Conference on Concrete Spalling due to Fire Exposure (2011) 453 pp., ISBN: 978-2-35158-118-6; e-ISBN: 978-2-35158-119-3; *Eds. E.A.B. Koenders and F. Dehn*

PRO 81: 2nd International RILEM Conference on Strain Hardening Cementitious Composites (SHCC2-Rio) (2011) 451 pp., ISBN: 978-2-35158-120-9; e-ISBN: 978-2-35158-121-6; *Eds. R.D. Toledo Filho, F.A. Silva, E.A.B. Koenders and E.M.R. Fairbairn*

PRO 82: 2nd International RILEM Conference on Progress of Recycling in the Built Environment (2011) 507 pp., e-ISBN: 978-2-35158-122-3; *Eds. V.M. John, E. Vazquez, S.C. Angulo and C. Ulsen*

PRO 83: 2nd International Conference on Microstructural-related Durability of Cementitious Composites (2012) 250 pp., ISBN: 978-2-35158-129-2; e-ISBN: 978-2-35158-123-0; *Eds. G. Ye, K. van Breugel, W. Sun and C. Miao*

PRO 84: CONSEC13—Seventh International Conference on Concrete under Severe Conditions—Environment and Loading (2013) 1930 pp., ISBN: 978-2-35158-124-7; e-ISBN: 978-2-35158-134-6; *Eds. Z.J. Li, W. Sun, C.W. Miao, K. Sakai, O.E. Gjorv and N. Banthia*

PRO 85: RILEM-JCI International Workshop on Crack Control of Mass Concrete and Related issues concerning Early-Age of Concrete Structures—ConCrack 3—Control of Cracking in Concrete Structures 3 (2012) 237 pp., ISBN: 978-2-35158-125-4; e-ISBN: 978-2-35158-126-1; *Eds. F. Toutlemonde and J.-M. Torrenti*

PRO 86: International Symposium on Life Cycle Assessment and Construction (2012) 414 pp., ISBN: 978-2-35158-127-8, e-ISBN: 978-2-35158-128-5; *Eds. A. Ventura and C. de la Roche*

PRO 87: UHPFRC 2013—RILEM-fib-AFGC International Symposium on Ultra-High Performance Fibre-Reinforced Concrete (2013), ISBN: 978-2-35158-130-8, e-ISBN: 978-2-35158-131-5; *Eds. F. Toutlemonde*

PRO 88: 8th RILEM International Symposium on Fibre Reinforced Concrete (2012) 344 pp., ISBN: 978-2-35158-132-2; e-ISBN: 978-2-35158-133-9; *Eds. Joaquim A.O. Barros*

PRO 89: RILEM International workshop on performance-based specification and control of concrete durability (2014) 678 pp., ISBN: 978-2-35158-135-3; e-ISBN: 978-2-35158-136-0; *Eds. D. Bjegović, H. Beushausen and M. Serdar*

PRO 90: 7th RILEM International Conference on Self-Compacting Concrete and of the 1st RILEM International Conference on Rheology and Processing of Construction Materials (2013) 396 pp., ISBN: 978-2-35158-137-7; e-ISBN: 978-2-35158-138-4; *Eds. Nicolas Roussel and Hela Bessaies-Bey*

PRO 91: CONMOD 2014—RILEM International Symposium on Concrete Modelling (2014), ISBN: 978-2-35158-139-1; e-ISBN: 978-2-35158-140-7; *Eds. Kefei Li, Peiyu Yan and Rongwei Yang*

PRO 92: CAM 2014—2nd International Conference on advances in chemically-activated materials (2014) 392 pp., ISBN: 978-2-35158-141-4; e-ISBN: 978-2-35158-142-1; *Eds. Caijun Shi and Xiadong Shen*

PRO 93: SCC 2014—3rd International Symposium on Design, Performance and Use of Self-Consolidating Concrete (2014) 438 pp., ISBN: 978-2-35158-143-8; e-ISBN: 978-2-35158-144-5; *Eds. Caijun Shi, Zhihua Ou and Kamal H. Khayat*

PRO 94 (online version): HPFRCC-7—7th RILEM conference on High performance fiber reinforced cement composites (2015), e-ISBN: 978-2-35158-146-9; *Eds. H.W. Reinhardt, G.J. Parra-Montesinos and H. Garrecht*

PRO 95: International RILEM Conference on Application of superabsorbent polymers and other new admixtures in concrete construction (2014), ISBN: 978-2-35158-147-6; e-ISBN: 978-2-35158-148-3; *Eds. Viktor Mechtcherine and Christof Schroefl*

PRO 96 (online version): XIII DBMC: XIII International Conference on Durability of Building Materials and Components (2015), e-ISBN: 978-2-35158-149-0; *Eds. M. Quattrone and V.M. John*

PRO 97: SHCC3—3rd International RILEM Conference on Strain Hardening Cementitious Composites (2014), ISBN: 978-2-35158-150-6; e-ISBN: 978-2-35158-151-3; *Eds. E. Schlangen, M.G. Sierra Beltran, M. Lukovic and G. Ye*

PRO 98: FERRO-11—11th International Symposium on Ferrocement and 3rd ICTRC—International Conference on Textile Reinforced Concrete (2015), ISBN: 978-2-35158-152-0; e-ISBN: 978-2-35158-153-7; *Ed. W. Brameshuber*

PRO 99 (online version): ICBBM 2015—1st International Conference on Bio-Based Building Materials (2015), e-ISBN: 978-2-35158-154-4; *Eds. S. Amziane and M. Sonebi*

PRO 100: SCC16—RILEM Self-Consolidating Concrete Conference (2016), ISBN: 978-2-35158-156-8; e-ISBN: 978-2-35158-157-5; *Ed. Kamal H. Kayat*

PRO 101 (online version): III Progress of Recycling in the Built Environment (2015), e-ISBN: 978-2-35158-158-2; *Eds I. Martins, C. Ulsen and S. C. Angulo*

PRO 102 (online version): RILEM Conference on Microorganisms-Cementitious Materials Interactions (2016), e-ISBN: 978-2-35158-160-5; *Eds. Alexandra Bertron, Henk Jonkers and Virginie Wiktor*

PRO 103 (online version): ACESC'16—Advances in Civil Engineering and Sustainable Construction (2016), e-ISBN: 978-2-35158-161-2; *Eds. T.Ch. Madhavi, G. Prabhakar, Santhosh Ram and P.M. Rameshwaran*

PRO 104 (online version): SSCS'2015—Numerical Modeling—Strategies for Sustainable Concrete Structures (2015), e-ISBN: 978-2-35158-162-9

PRO 105: 1st International Conference on UHPC Materials and Structures (2016), ISBN: 978-2-35158-164-3; e-ISBN: 978-2-35158-165-0

PRO 106: AFGC-ACI-fib-RILEM International Conference on Ultra-High-Performance Fibre-Reinforced Concrete—UHPFRC 2017 (2017), ISBN: 978-2-35158-166-7; e-ISBN: 978-2-35158-167-4; *Eds. François Toutlemonde and Jacques Resplendino*

PRO 107 (online version): XIV DBMC—14th International Conference on Durability of Building Materials and Components (2017), e-ISBN: 978-2-35158-159-9; *Eds. Geert De Schutter, Nele De Belie, Arnold Janssens and Nathan Van Den Bossche*

PRO 108: MSSCE 2016—Innovation of Teaching in Materials and Structures (2016), ISBN: 978-2-35158-178-0; e-ISBN: 978-2-35158-179-7; *Ed. Per Goltermann*

PRO 109 (2 volumes): MSSCE 2016—Service Life of Cement-Based Materials and Structures (2016), ISBN Vol. 1: 978-2-35158-170-4; Vol. 2: 978-2-35158-171-4; Set Vol. 1&2: 978-2-35158-172-8; e-ISBN: 978-2-35158-173-5; *Eds. Miguel Azenha, Ivan Gabrijel, Dirk Schlicke, Terje Kanstad and Ole Mejlhede Jensen*

PRO 110: MSSCE 2016—Historical Masonry (2016), ISBN: 978-2-35158-178-0; e-ISBN: 978-2-35158-179-7; *Eds. Inge Rörig-Dalgaard and Ioannis Ioannou*

PRO 111: MSSCE 2016—Electrochemistry in Civil Engineering (2016); ISBN: 978-2-35158-176-6; e-ISBN: 978-2-35158-177-3; *Ed. Lisbeth M. Ottosen*

PRO 112: MSSCE 2016—Moisture in Materials and Structures (2016), ISBN: 978-2-35158-178-0; e-ISBN: 978-2-35158-179-7; *Eds. Kurt Kielsgaard Hansen, Carsten Rode and Lars-Olof Nilsson*

PRO 113: MSSCE 2016—Concrete with Supplementary Cementitious Materials (2016), ISBN: 978-2-35158-178-0; e-ISBN: 978-2-35158-179-7; *Eds. Ole Mejlhede Jensen, Konstantin Kovler and Nele De Belie*

PRO 114: MSSCE 2016—Frost Action in Concrete (2016), ISBN: 978-2-35158-182-7; e-ISBN: 978-2-35158-183-4; *Eds. Marianne Tange Hasholt, Katja Fridh and R. Doug Hooton*

PRO 115: MSSCE 2016—Fresh Concrete (2016), ISBN: 978-2-35158-184-1; e-ISBN: 978-2-35158-185-8; *Eds. Lars N. Thrane, Claus Pade, Oldrich Svec and Nicolas Roussel*

PRO 116: BEFIB 2016—9th RILEM International Symposium on Fiber Reinforced Concrete (2016), ISBN: 978-2-35158-187-2; e-ISBN: 978-2-35158-186-5; *Eds. N. Banthia, M. di Prisco and S. Soleimani-Dashtaki*

PRO 117: 3rd International RILEM Conference on Microstructure Related Durability of Cementitious Composites (2016), ISBN: 978-2-35158-188-9; e-ISBN: 978-2-35158-189-6; *Eds. Changwen Miao, Wei Sun, Jiaping Liu, Huisu Chen, Guang Ye and Klaas van Breugel*

PRO 118 (4 volumes): International Conference on Advances in Construction Materials and Systems (2017), ISBN Set: 978-2-35158-190-2; Vol. 1: 978-2-35158-193-3; Vol. 2: 978-2-35158-194-0; Vol. 3: ISBN:978-2-35158-195-7; Vol. 4: ISBN:978-2-35158-196-4; e-ISBN: 978-2-35158-191-9; *Ed. Manu Santhanam*

PRO 119 (online version): ICBBM 2017—Second International RILEM Conference on Bio-based Building Materials, (2017), e-ISBN: 978-2-35158-192-6; *Ed. Sofiane Amziane*

PRO 120 (2 volumes): EAC-02—2nd International RILEM/COST Conference on Early Age Cracking and Serviceability in Cement-based Materials and Structures, (2017), Vol. 1: 978-2-35158-199-5, Vol. 2: 978-2-35158-200-8, Set: 978-2-35158-197-1, e-ISBN: 978-2-35158-198-8; *Eds. Stéphanie Staquet and Dimitrios Aggelis*

PRO 121 (2 volumes): SynerCrete18: Interdisciplinary Approaches for Cement-based Materials and Structural Concrete: Synergizing Expertise and Bridging Scales of Space and Time, (2018), Set: 978-2-35158-202-2, Vol.1: 978-2-35158-211-4, Vol.2: 978-2-35158-212-1, e-ISBN: 978-2-35158-203-9; *Eds. Miguel Azenha, Dirk Schlicke, Farid Benboudjema, Agnieszka Knoppik*

PRO 122: SCC'2018 China—Fourth International Symposium on Design, Performance and Use of Self-Consolidating Concrete, (2018), ISBN: 978-2-35158-204-6, e-ISBN: 978-2-35158-205-3; *Eds. C. Shi, Z. Zhang, K. H. Khayat*

PRO 123: Final Conference of RILEM TC 253-MCI: Microorganisms-Cementitious Materials Interactions (2018), Set: 978-2-35158-207-7, Vol.1: 978-2-35158-209-1, Vol.2: 978-2-35158-210-7, e-ISBN: 978-2-35158-206-0; *Ed. Alexandra Bertron*

PRO 124 (online version): Fourth International Conference Progress of Recycling in the Built Environment (2018), e-ISBN: 978-2-35158-208-4; *Eds. Isabel M. Martins, Carina Ulsen, Yury Villagran*

PRO 125 (online version): SLD4—4th International Conference on Service Life Design for Infrastructures (2018), e-ISBN: 978-2-35158-213-8; *Eds. Guang Ye, Yong Yuan, Claudia Romero Rodriguez, Hongzhi Zhang, Branko Savija*

PRO 126: Workshop on Concrete Modelling and Material Behaviour in honor of Professor Klaas van Breugel (2018), ISBN: 978-2-35158-214-5, e-ISBN: 978-2-35158-215-2; *Ed. Guang Ye*

PRO 127 (online version): CONMOD2018—Symposium on Concrete Modelling (2018), e-ISBN: 978-2-35158-216-9; *Eds. Erik Schlangen, Geert de Schutter, Branko Savija, Hongzhi Zhang, Claudia Romero Rodriguez*

PRO 128: SMSS2019—International Conference on Sustainable Materials, Systems and Structures (2019), ISBN: 978-2-35158-217-6, e-ISBN: 978-2-35158-218-3

PRO 129: 2nd International Conference on UHPC Materials and Structures (UHPC2018-China), ISBN: 978-2-35158-219-0, e-ISBN: 978-2-35158-220-6

PRO 130: 5th Historic Mortars Conference (2019), ISBN: 978-2-35158-221-3, e-ISBN: 978-2-35158-222-0; *Eds. José Ignacio Álvarez, José María Fernández, Íñigo Navarro, Adrián Durán, Rafael Sirera*

PRO 131 (online version): 3rd International Conference on Bio-Based Building Materials (ICBBM2019), e-ISBN: 978-2-35158-229-9; *Eds. Mohammed Sonebi, Sofiane Amziane, Jonathan Page*

PRO 132: IRWRMC'18—International RILEM Workshop on Rheological Measurements of Cement-based Materials (2018), ISBN: 978-2-35158-230-5, e-ISBN: 978-2-35158-231-2; *Eds. Chafika Djelal, Yannick Vanhove*

PRO 133 (online version): CO2STO2019—International Workshop CO2 Storage in Concrete (2019), e-ISBN: 978-2-35158-232-9; *Eds. Assia Djerbi, Othman Omikrine-Metalssi, Teddy Fen-Chong*

PRO 134: 3rd ACF/HNU International Conference on UHPC Materials and Structures—UHPC'2020, ISBN: 978-2-35158-233-6, e-ISBN: 978-2-35158-234-3; *Eds. Caijun Shi and Jiaping Liu*

PRO 135: Fourth International Conference on Chemically Activated Materials (CAM2021), ISBN: 978-2-35158-235-0, e-ISBN: 978-2-35158-236-7; *Eds. Caijun Shi and Xiang Hu*

PRO 136: Lime Based Materials for Repairing Historic Structures (2022), ISBN: 978-2-35158-240-4, e-ISBN: 978-2-35158-237-4; *Eds. Ioanna Papayianni and Jan Válek*

PRO 137 (online version): V International Conference Progress of Recycling in the Built Environment (2023), e-ISBN: 978-2-35158-238-1; *Eds. Ines Döring, Yury Villagrán Zaccardi, Anette Müller*

RILEM Reports (REP)

Report 19: Considerations for Use in Managing the Aging of Nuclear Power Plant Concrete Structures (ISBN: 2-912143-07-1); *Ed. D. J. Naus*

Report 20: Engineering and Transport Properties of the Interfacial Transition Zone in Cementitious Composites (ISBN: 2-912143-08-X); *Eds. M. G. Alexander, G. Arliguie, G. Ballivy, A. Bentur and J. Marchand*

Report 21: Durability of Building Sealants (ISBN: 2-912143-12-8); *Ed. A. T. Wolf*

Report 22: Sustainable Raw Materials—Construction and Demolition Waste (ISBN: 2-912143-17-9); *Eds. C. F. Hendriks and H. S. Pietersen*

Report 23: Self-Compacting Concrete state-of-the-art report (ISBN: 2-912143-23-3); *Eds. Å. Skarendahl and Ö. Petersson*

Report 24: Workability and Rheology of Fresh Concrete: Compendium of Tests (ISBN: 2-912143-32-2); *Eds. P. J. M. Bartos, M. Sonebi and A. K. Tamimi*

Report 25: Early Age Cracking in Cementitious Systems (ISBN: 2-912143-33-0); *Ed. A. Bentur*

Report 26: Towards Sustainable Roofing (Joint Committee CIB/RILEM) (CD 07) (e-ISBN 978-2-912143-65-5); *Eds. Thomas W. Hutchinson and Keith Roberts*

Report 27: Condition Assessment of Roofs (Joint Committee CIB/RILEM) (CD 08) (e-ISBN 978-2-912143-66-2); *Ed. CIB W 83/RILEM TC166-RMS*

Report 28: Final report of RILEM TC 167-COM 'Characterisation of Old Mortars with Respect to Their Repair (ISBN: 978-2-912143-56-3); *Eds. C. Groot, G. Ashall and J. Hughes*

Report 29: Pavement Performance Prediction and Evaluation (PPPE): Interlaboratory Tests (e-ISBN: 2-912143-68-3); *Eds. M. Partl and H. Piber*

Report 30: Final Report of RILEM TC 198-URM 'Use of Recycled Materials' (ISBN: 2-912143-82-9; e-ISBN: 2-912143-69-1); *Eds. Ch. F. Hendriks, G. M. T. Janssen and E. Vázquez*

Report 31: Final Report of RILEM TC 185-ATC 'Advanced testing of cement-based materials during setting and hardening' (ISBN: 2-912143-81-0; e-ISBN: 2-912143-70-5); *Eds. H. W. Reinhardt and C. U. Grosse*

Report 32: Probabilistic Assessment of Existing Structures. A JCSS publication (ISBN 2-912143-24-1); *Ed. D. Diamantidis*

Report 33: State-of-the-Art Report of RILEM Technical Committee TC 184-IFE 'Industrial Floors' (ISBN 2-35158-006-0); *Ed. P. Seidler*

Report 34: Report of RILEM Technical Committee TC 147-FMB 'Fracture mechanics applications to anchorage and bond' Tension of Reinforced Concrete Prisms—Round Robin Analysis and Tests on Bond (e-ISBN 2-912143-91-8); *Eds. L. Elfgren and K. Noghabai*

Report 35: Final Report of RILEM Technical Committee TC 188-CSC 'Casting of Self Compacting Concrete' (ISBN 2-35158-001-X; e-ISBN: 2-912143-98-5); *Eds. Å. Skarendahl and P. Billberg*

Report 36: State-of-the-Art Report of RILEM Technical Committee TC 201-TRC 'Textile Reinforced Concrete' (ISBN 2-912143-99-3); *Ed. W. Brameshuber*

Report 37: State-of-the-Art Report of RILEM Technical Committee TC 192-ECM 'Environment-conscious construction materials and systems' (ISBN: 978-2-35158-053-0); *Eds. N. Kashino, D. Van Gemert and K. Imamoto*

Report 38: State-of-the-Art Report of RILEM Technical Committee TC 205-DSC 'Durability of Self-Compacting Concrete' (ISBN: 978-2-35158-048-6); *Eds. G. De Schutter and K. Audenaert*

Report 39: Final Report of RILEM Technical Committee TC 187-SOC 'Experimental determination of the stress-crack opening curve for concrete in tension' (ISBN 978-2-35158-049-3); *Ed. J. Planas*

Report 40: State-of-the-Art Report of RILEM Technical Committee TC 189-NEC 'Non-Destructive Evaluation of the Penetrability and Thickness of the Concrete Cover' (ISBN 978-2-35158-054-7); *Eds. R. Torrent and L. Fernández Luco*

Report 41: State-of-the-Art Report of RILEM Technical Committee TC 196-ICC 'Internal Curing of Concrete' (ISBN 978-2-35158-009-7); *Eds. K. Kovler and O. M. Jensen*

Report 42: 'Acoustic Emission and Related Non-destructive Evaluation Techniques for Crack Detection and Damage Evaluation in Concrete'—Final Report of RILEM Technical Committee 212-ACD (e-ISBN: 978-2-35158-100-1); *Ed. M. Ohtsu*

Report 45: Repair Mortars for Historic Masonry—State-of-the-Art Report of RILEM Technical Committee TC 203-RHM (e-ISBN: 978-2-35158-163-6); *Eds. Paul Maurenbrecher and Caspar Groot*

Report 46: Surface delamination of concrete industrial ffioors and other durability related aspects guide—Report of RILEM Technical Committee TC 268-SIF (e-ISBN: 978-2-35158-201-5); *Ed. Valerie Pollet*

Contents

Acronyms

AC	Asphalt concrete
ARA	Asphalt reuse/recycling agent
ATR	Attenuated total reflection
BBR	Bending beam rheometer
BSM	Bitumen stabilized materials
CAM	Cold asphalt mixtures
CBTM	Cement-bitumen treated materials
CBTM-EMU	Cement-bitumen treated materials (bitumen emulsion)
CBTM-FB	Cement-bitumen treated materials (foamed bitumen)
CCPR	Cold central plant recycling
CIR	Cold in-place recycling
CRM	Cold recycled material
CTM	Cement treated material
DSR	Dynamic shear rheometer
ER_m	Expansion ratio
FDR	Full depth reclamation
FSD	Free surface dry
FT	Fuller-Thompson
FTIR	Fourier transform infrared spectroscopy
HMA	Hot-mix asphalt
ILTP	Inter-laboratory test program
ITS	Indirect tensile strength
ITSM	Indirect tensile stiffness modulus
LTA	Long-term ageing
PAV	Pressure aging vessel
PCS	Passing a control Sieve
Pen	Penetration value
PSD	Partially surface dry
RA	Reclaimed asphalt
RAb	Reclaimed asphalt binder
RAP	Reclaimed asphalt pavement

RBSP	Ring and ball softening point temperature
RRT	Round-robin test
RTFOT	Rolling thin film oven test
SGC	Shear gyratory compactor
SSD	Saturated surface dry
STA	Short-term ageing
Vb	Virgin binder
VFB	Voids filled with bitumen
VFL	Voids Filled with liquid
VMA	Voids in mineral aggregates

Chapter 1
Introduction

Gabriele Tebaldi and Eshan V. Dave

Keywords Reclaimed asphalt · Asphalt recycling · Binder interaction · Additives · Mix design · Life-cycle assessment · Sustainable pavements

1.1 Background and Motivation

The use of recycled asphalt (RA) in pavement mixtures has been on continuous increase. This is driven by combination of economic, environmental and social reasons. The sources of new materials for asphalt manufacture are rapidly diminishing, for example fewer and fewer aggregate mining permits are being issued. The currently used asphalt mixture design methods are based on extension of methods that were originally developed for proportioning of new materials. Furthermore, majority of these methods are based on the empirical information and the approaches for mixtures with RA have been developed for relatively lower RA contents (typically up to 30%). With rise in use of cold as well non-cold recycled asphalt, there is an urgent need to gain fundamental understanding of interactions between various asphalt mixture constituents, specifically the interaction of RA and new constituents (including additives such as rejuvenators). The current needs are in the area of mix design standardization, best practices manuals and life-cycle assessment tools to aid in correct selection of materials, methodology and proportioning of RA mixtures.

G. Tebaldi (✉)
Department of Engineering and Architecture, University of Parma, Parma, Italy
e-mail: gtebaldi@unipr.it

E. V. Dave (✉)
Department of Civil and Environmental Engineering, University of New Hampshire, Durham, NH, USA
e-mail: Eshan.Dave@unh.edu

G. Tebaldi and E. Dave (eds.), *Recycling Asphalt Pavement*, RILEM State-of-the-Art Reports 44, https://doi.org/10.1007/978-3-031-94231-0_1

Motivated by the lack of knowledge and empirical approaches currently used in engineering practice, the RILEM TC 264-RAP undertook interlaboratory studies, organized industry workshops, survey of professionals and analysis of test data. Outcome of these activities resulted in this STAR that attempts to fulfills scientific knowledge gaps and serves as a reference for practitioners. The main contribution of this state of the art report is to enhance the fundamental understanding on the interactions between recycled asphalt (RA) materials and new virgin materials in asphalt paving mixtures.

1.2 Scope of this Book

The scope of this STAR is limited to RA and usage of RA in hot, warm and cold asphalt mixtures. Specifically, the range of activities and outcomes are as summarized below:

- Cold asphalt recycling discussions are focused on laboratory mix design, specimen preparation, and curing methods associated with use of cement as active filler in bituminous stabilized mixtures. This is discussed in Chap. 2 of this STAR that was developed by the task group 1 of the TC.
- Use of RA in warm mixed of asphalt mixtures in terms of mechanical properties and performance indicators was explored through interlaboratory studies by the task group 2 of the TC. These results and findings are discussed in Chap. 3 of this STAR.
- Rejuvenators as additives to support increased performance as well as possibility to use larger amounts of RA in asphalt mixtures was explored by task group 3 of the TC through assessment of bitumen blends (blends of virgin bitumen, bitumen from RA and rejuvenators). Specifically, laboratory aging procedures and rheological as well as chemical characterization was conducted through round-robin and interlaboratory testing campaigns. Experimentation details, results and findings are discussed in Chap. 4.
- Defining and evaluating the basis that can be used to quantify the availability of bitumen from RA to asphalt mixtures was undertaken by task group 5 of the TC. Through interlaboratory testing campaign, a preliminary protocol was developed for determining degree of binder availability. Chapter 5 of this STAR discussing results and outcomes of this effort.
- Detailed literature review and results from survey of professionals as well as outcomes of an international webinar enabled task group 4 of the TC to identify the current knowledge gaps for application of life-cycle assessment tools to asphalt mixtures with RAs. The knowledge gaps as well as necessary next steps to fill these gaps with respect to use of LCA for asphalt mixtures with RAs are discussed within the last chapter of this STAR.

Detailed literature review and results from survey of professionals as Finally, the overall efforts the TC activities that are dis-cussed in this STAR are summarized, key conclusions are highlighted and the needs for future collaborative research activities are presented in Chap. 6.

Chapter 2
Impact of the Compaction and Curing on the Properties of Cement-Bitumen Treated Materials (CBTM)

Daniel Perraton, Alan Carter, Simone Raschia, Andrea Grilli, Andrea Graziani, Emiliano Pasquini, Patricia Kara De Maeijer, Davide Lo Presti, Marco Pasetto, Gabriele Tebaldi, and Eshan Dave

D. Perraton (✉) · A. Carter · S. Raschia
Department of Construction Engineering, École de Technologie Supérieure (LCMB-ETS), Montreal, QC, Canada
e-mail: daniel.perraton@etsmtl.ca

A. Carter
e-mail: alan.carter@etsmtl.ca

S. Raschia
e-mail: s.raschia@stsmobile.it

A. Grilli
Department of Economics, Science and Law, University of the Republic of San Marino, San Marino City, San Marino
e-mail: andrea.grilli@unirsm.sm

A. Graziani
Department of Civil and Building Engineering and Architecture, Marche Polytechnic University, Ancona, Italy
e-mail: a.graziani@univpm.it

E. Pasquini · M. Pasetto
Department of Civil, Environmental and Architectural Engineering (ICEA), University of Padua, Padua, Italy
e-mail: emiliano.pasquini@unipd.it

M. Pasetto
e-mail: marco.pasetto@unipd.it

P. K. De Maeijer
EMIB, Faculty of Applied Engineering, University of Antwerp, Antwerp, Belgium
e-mail: patricija.karademaeijer@uantwerpen.be

G. Tebaldi and E. Dave (eds.), *Recycling Asphalt Pavement*, RILEM State-of-the-Art Reports 44, https://doi.org/10.1007/978-3-031-94231-0_2

Keywords Cold recycling materials-CRM · Cement-Bitumen treated materials-CBTM · Curing · Compaction · Reclaimed asphalt-RA · Emulsion · Foam · Volumetric approach · Strength and stiffness

2.1 Introduction

Recycling of asphalt pavements leads to substantial savings in energy and materials. In particular, cold recycling technologies, based on the use of bitumen emulsion or foamed bitumen, allow the production of bituminous mixtures at ambient temperature, and with wet aggregates. To promote a proper distribution of the bitumen and to manage the water inside of the mixture, active filler as Portland cement, hydrated lime, fly ash, etc., are often incorporated in the formulation.

2.1.1 Quality Control of RA: How to Deal with It?

The use of processed materials coming from the milling or demolition of aged pavements to reconstruct/rehabilitate road structures allow matching technical and performance requirements while preserving natural resources. Materials derived from the milling or demolition of asphalt layers or from surplus or rejection of asphalt concrete (AC) production are generally called *Reclaimed Asphalt* (RA) or *Reclaimed Asphalt Pavement* (RAP) [1]. Currently, there are multiple methods to recycle these materials: cold recycling, warm recycling, half-warm recycling and hot recycling. State-of-the-Art Report of the RILEM Technical Committee 237-SIB [2], has clearly shown the lack of RA characterization methods to identify, to rank the RA fractions, and able to identify which influence the properties of CRM mixtures. Moreover, a wide discrepancy amongst the methodologies utilized for specimen preparation and curing procedures was highlighted. In this context, the material characterization, the mix

D. L. Presti
Department of Engineering, University of Palermo, Palermo, Italy

Nottingham Transportation Engineering Centre, University of Nottingham, Nottingham, UK

D. L. Presti
e-mail: davide.lopresti@you.unipa.it; davide.lopresti@nottingham.ac.uk

G. Tebaldi (✉)
Department of Engineering and Architecture, University of Parma, Parma, Italy
e-mail: gabriele.tebaldi@unipr.it

E. Dave (✉)
Department of Civil and Environmental Engineering, University of New Hampshire, Durham, NH, USA
e-mail: eshan.dave@unh.edu

design and the curing process are key phases that still need research to adapt and/or modify scientific and technical practices. The development of standard methods based on scientific and technical concepts will contribute to a better understanding and sharing of knowledge and to promote the use of CRM.

A round-robin test (RRT) performed by the RILEM TC 237-SIB TG6 [2] allowed to point out that new test methods could be used to quickly characterize different sources of RA materials. Two tests were investigated: a fragmentation test and a cohesion test. Results of the RRT program showed the capability of both tests to characterize and to classify RA in accordance with the processing and source.

2.2 Objective and Research Approach

Based on the reflection started up by the RILEM TC 237-SIB TG6, the main objective of the RILEM TC 264-RAP TG1 is to better understand the impact of different aspects related to CRM mix design on their properties. Based on an inter-laboratory test program (ILTP), a multi-country research project has been launched to:

a. Evaluate the ability of the fragmentation test to highlight differences in the properties of CRM with different RA source;
b. Quantify the impact of the compaction method used for laboratory specimen preparation (Marshall, Proctor and shear gyratory compactors);
c. Understand the influence of different types of laboratory curing (free surface drying-FSD; partially surface drying-PSD);
d. Evaluate the effect of the type of bituminous binder (bitumen emulsion and foam).

The result analysis was implemented with a volumetric approach to quantify constituent proportions of the CRM based on the same concept already used for hot-mix asphalt (HMA) mixtures.

2.3 Cold Recycling

Before starting the ILTP, it was essential to do a literature review of the work done on CRM around the world. The review focused on the definition of CRM, on the mix design and on the laboratory procedures.

2.3.1 Techniques and Mixtures

Cold recycling can be applied in new construction or more commonly in the rehabilitation of deteriorated asphalt pavement layers with the partial or total re-use of RA. CRM mixtures do not need addition of heat to increase the temperature of the aggregates or to dry them, there is only a need of a limited amount of heat to keep stable the emulsion and to make fluid the bitumen for foaming.

Cold recycling can be performed either in a production plant, Cold Central Plant Recycling (CCPR), or in-place, Cold In-place Recycling (CIR). CCPR technique employs an RA aggregate obtained from the processing of RA (crushing and screening) performed at the central plant. In this way, the producer keeps the variability of the RA aggregate and the overall gradation of the mixture under control, which improves the quality and the stability of the final CRM mixture [3].

The in-plant production is often performed with mobile facilities that can be installed near or at the job site. This equipment does not allow heating and aggregates sieving. An accurate proportioning of the input materials, as well as a thorough preliminary quality control, increase the confidence to obtain mixtures that meet the requirements [4].

After hauling to the jobsite, CCPR mixtures are used to build cold recycled layers with thicknesses between 10 and 25 cm by means of the same equipment used for HMA. Pavers are used to laydown the CRM layer and control its thickness and slope. The compaction is carried out by means of vibrating smooth drums and pneumatic tire rollers. The required level of compaction can be achieved more easily by the selection of an appropriate aggregate gradation, water dosage and layer thickness. CCPR mixtures can be used for binder and/or base courses, as a high confidence on the mixtures stability and performance can be obtained [5].

In the current practice, CIR indicates partial-depth recycling including only HMA layers. With CIR technique, milling, mixing and compaction are simultaneously performed on-site by using single-unit or multi-unit trains. Therefore, the main advantage with respect to CCPR is that CIR does not require material transportation (or just limited amount considering water, binders, and aggregates to correct the gradation of the in-situ material to be recycled). A recycling train is normally equipped with volumetric containers for spreading active filler, a recycler coupled to a tank truck for the addition of water, a tank truck for the bituminous binder (emulsion or bitumen for foaming), a vibrating smooth drum roller, a pneumatic tire roller and a grader for shaping and levelling. Recycling machines (recyclers), which combine milling and mixing phases, operate in a single or more passes depending on the required fineness or homogeneity level to be achieved. The use of CIR leads to significant economic and environmental advantages because of the elimination of the haulage. However, in this process material control is more difficult and thus is demanding to obtain high-quality mixtures [6].

When recycling involves both the HMA and the underlying layers, the term full-depth reclamation (FDR) is used [7]. In FDR, similar to CIR, virgin aggregates (including mineral filler) can be added to meet gradation requirements. Due to their high heterogeneity FDR mixtures are generally used to build foundation courses, or base courses when low traffic volume is expected [8].

In cold recycling techniques, the use of bitumen emulsion or foamed bitumen, which employ water in different ways, to allow the distribution of the bitumen in the mixture without act on temperature to reduce the viscosity of the binder, allows high amount of RA at ambient temperature to be recycled. Normally, mineral additions

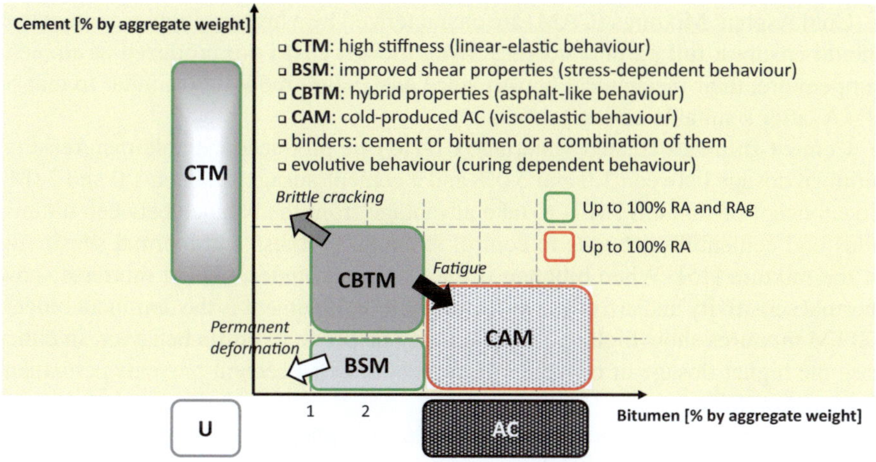

Fig. 2.1 Conceptual composition of mixtures used in road construction

like cement, lime, fly ash, etc., are used to enhance cohesion and bitumen dispersion, as well as to increase the initial rate of the curing process and to improve short- and long-term mechanical properties. Often the term "active fillers" is used to indicate those mineral additions [9]. It should be highlighted that, when emulsion is use "conventional fillers" like limestone dust may be extremely reactive and affect the braking and setting behavior of the CRM [10].

The mechanical behavior of CRM mixtures may vary significantly by changing the aggregate quality and composition, as well as the type and dosage of stabilizing agents (bitumen emulsion, foamed bitumen, cement, etc.) [11]. Figure 2.1 shows an attempt to classify pavement mixtures based on the content of cement and bitumen. It is remarked that this classification includes mixtures produced with or without recycled aggregate, and that the indicated composition limits are not strict.

Along the y-axis we find the Cement Treated Materials (CTM) that are normally produced with a cement dosage ranging from 1.5 to 6.0%. They show remarkable compressive strength, stiffness and bearing capacity, and their brittleness increases as the cement content increases. CTM can be manufactured both in plant and in situ, using the FDR technique [12].

Bitumen Stabilized Materials (BSM) are characterized by a residual bitumen content up to 3% (bitumen residual of emulsion or foamed bitumen) [13] and by an amount of active filler determined in function of the mechanical performance in the mix design process; the most common active filler is the Portland cement, in this case the content is up to 1.0%. Bitumen is the main binder, whereas cement is used to improve the short-term performance. BSM were developed to improve the mechanical behavior of unbound materials. In fact, they are characterized by a stress-dependent behavior and enhanced shear resistance properties. The primary mode of failure of BSMs is permanent deformation [14].

Cold Asphalt Mixtures (CAM) are characterized by a higher bitumen content that should ensure a full particle coating. Even though CAM are produced at ambient temperature, their mechanical performance and failure modes are similar to that of HMA, after a suitable curing process.

Cement-Bitumen Treated Materials (CBTM) are produced with bitumen/residual bitumen dosage between 1.0 and 3.0% and a cement dosage between 1.0 and 3.0%. The concept behind CBTM is to take advantage from the balance between bituminous and cementitious bonds to control strength, stiffness and thermal sensitivity of the mixture [15]. When bitumen is the dominant binder, CBTM mixtures show thermal sensitivity and are prone to fatigue failure. If cement is the dominant binder, CBTM mixtures show high stiffness and may be prone to brittle behavior. In either case the higher dosage of cement with respect to BSM cement prevents permanent deformation failure.

In the pavement structure, when CBTM layer is placed between HMA and CTM or cement-stabilized soils, it acts as a transition layer between a flexible one and a rigid/bearing one. In these cases, a stiff foundation layer facilitates the effectiveness of the compaction phase of CBTM.

2.3.2 Constituent Materials

2.3.2.1 Reclaimed Asphalt

Reclaimed asphalt is obtained from the pavement milling process, after which is stockpiled in plants to be re-employed in totally or partially recycled materials. RA is characterized by an advanced stage of oxidation and ageing, achieved during the service life and promoted by air and environmental factors (air, humidity, temperature, traffic, water, etc.).

RA content employed in HMA usually goes from 10 to 50%, and it depends on the traffic level (high, medium, low), the type of material (open graded, dense graded, gap graded) and the pavement layer considered (surface course, binder course, base course). When using RA, to guarantee an optimal degree of workability and RA binder reactivation, attention should be paid to the type and dosage of recycling additives.

On the other hand, when RA is employed to produce cold recycled layers, in particular CBTM materials, it is possible to re-use up to 100% of RA in the aggregate gradation [16]. In this context recycling additives are not used, and because of the ambient production temperatures, RA aged binder does not reactivate and blend with the virgin binder (bitumen emulsion or foamed bitumen).

One issue related to the use of RA in CBTM materials is the formation of aggregate clusters [17]. The presence of agglomerations is an important factor affecting the material gradation, hence its mechanical properties. In the framework of the RILEM TC 237-SIB TG6 [17] a round robin test was proposed to validate a protocol, the fragmentation test, to characterize different RA sources. The test is able to differentiate different RA materials in terms of crushing resistance and subsequent change of gradation, but more studies were recommended to correlate the fragmentation test classification with the properties of CBTM materials produced [18, 19].

An important part of the experimental plan developed in the TG1 research program was dedicated to the investigation of the effect of two RA sources on CBTM mechanical properties. The two RA materials were chosen according to the fragmentation test classification.

2.3.2.2 Bitumen Emulsions

Bitumen emulsions are obtained by the dispersion of bitumen droplets in an aqueous phase. Bitumen droplets in the bitumen emulsion are characterized by a size distribution, and this distribution is influenced by the bitumen emulsion recipe and the mechanics and operating conditions of the emulsion manufacturing plant. Water and bitumen remain separate phases and the stability of bitumen droplets is ensured by the presence of emulsifiers. Emulsifiers are often supplied in a water-insoluble form and need to be "activated" (or de-protonated) with acid or alkali to have the final surface-active cationic or anionic water-soluble form used to prepare the soap solution. The choice of the acid or alkali (and the final emulsion pH, subsequently), strongly influence its properties [20]. Although a large amount of emulsifier in an emulsion resides at the bitumen/water interface, a certain amount is dispersed in the water phase (free part). Bitumen emulsions normally consist of 40–75% of bitumen, 0.1–2.5% of emulsifier and 25–60% of water, plus secondary components [21].

2.3.2.3 Foam Bitumen

Foamed bitumen refers to a physical state of the binder and it is produced by injecting cold water into hot bitumen, resulting in spontaneous foaming. The physical properties of the bitumen are temporarily altered when the injected water, on contact with the hot bitumen, is turned into steam, which is trapped in thousands of tiny bitumen bubbles which generally dissipate in less than a minute.

To produce a CBTM with this technique, the bitumen is foamed on site and incorporated into the aggregate during mixing in that context while still in its foamed state. The greater the volume of the foam, the better the distribution of the bitumen into the aggregates' skeleton; the longer the foamed state, the better the CBTM workability.

The properties of the foam are characterized by means of the expansion ratio and the half-life values. These values are calculated as follows [22]:

- Expansion Ratio (ER_m) = Maximum volume of foamed bitumen/Original volume of bitumen;

Table 2.1 Properties of mineral fillers used in cold recycling

Parameter	Standard	Limestone filler	Dolomite filler
Passing at 2 mm sieve size [%]	EN 933-1 [24]	100	100
Passing at 0.125 mm sieve size [%]	EN 933-1 [24]	85–100	85–100
Passing at 0.063 mm sieve size [%]	EN 933-1 [24]	70–100	70–100
Density-ρ [g/cm^3]	EN 1097-7 [25]	2.450–2.750	2.650–2.850
Methylene blue [g/kg]	EN 933-9 [26]	<5	<5
Fineness (Blaine) [cm^2/g]	EN 196-6 [27]	>5700	>5700
Voids of dry compacted filler (Rigden) [%]	EN 1097-4 [28]	28–38	28–38
Stiffening power ΔR&B [°C]	EN 13179-1 [29]	8–16	8–16
Calcium carbonate $CaCO_3$ [%]	EN 196-21 [30]	>90	–

- Half-life ($\tau_{1/2}$) = Time measured in seconds for the foamed bitumen to subside from the maximum volume to half of the maximum volume.
- Standard specifications for ER_m and $\tau_{1/2}$ are non-existent but recommended values of $ER_m > 8$ and $\tau_{1/2} > 6$ s are observed from South African experience [23].

2.3.2.4 Mineral Filler

Mineral filler is often necessary by specifications for materials used in cold recycled layers to satisfy grading requirements since the RA is generally characterized by an inadequate content of fines due to agglomerations. Table 2.1 reports typical requirements/characteristics of two types of mineral fillers used in CBTM, being limestone, the type of filler mostly used.

2.3.2.5 Cement

In general, the addition of cement increases strength, stiffness and permanent deformation resistance of CBTM while negatively affecting its cracking resistance, in particular at high cement content [15, 31]. However, it is worth specifying that the mechanical response of a CBTM is extremely dependent on the proportions in which bitumen and cement are mixed. In this sense, Grilli et al. [15] clarified the behaviors of cement and bitumen acting at the same time by considering the weight ratio between the residual emulsion bitumen in the mix (B_R) and the cement dosage used (C). The cement allows high stiffness to be reached and, increasing the residual bitumen dosage (B_R), an increasing of fracture resistance can be provided, too.

As far as the interactions between cement and bitumen emulsion concern, besides the effects on the mechanical properties cited above, the cement can accelerate the emulsion breaking. Moreover, hydration products can be dispersed within the binder and could penetrate the bond between the binder and the aged bitumen of RA leading

Table 2.2 Properties of cements used in cold recycling

Parameter	Standard	CEM II/ B-LL 42.5R	CEM II/ B-LL 32.5R	CEM III/B 42.5 N
Strength after 2 days [MPa]	EN 196-1 [32]	≥20	≥10	≥20
Strength after 28 days [MPa]	EN 196-1 [32]	42.5–62.5	32.5–52.5	42.5–62.5
Fineness (Blaine) [cm^2/g]	EN 196-6 [27]	≥3300	≥4000	≥3500
Setting time [min]	EN 196-3 [33]	≥120	≥130	≥120
Density-ρ_f [kg/m^3]	EN 1097-7 [25]	2750–3250	2650–3150	2650–3150

to an additional effect on binder stiffening. Table 2.2 shows typical requirements of cements used in CBTM.

2.3.2.6 Hydrated Lime

Lime has been widely used as an additive or an active filler in bituminous mixtures to improve durability and likewise hydrated lime, either as a powder or as a slurry. It is typically used in CRM materials to improve the distribution of the bitumen in the mixture thanks to is high porosity and as a dewatering agent, since it accelerates emulsion breaking and curing process. Some experiments based on Zeta potential measurements were used to evaluate electrical change as an important factor which could influence the breaking characteristics of emulsions. Niazi and Jalili [34] have shown that CRM materials containing lime slurry have a superior moisture resistance compared to the mixture manufactured with lime powder. It seems that the balanced distribution of lime in the form of slurry might help the adhesion mechanism within the mixture. Properties of lime used in cold recycling are shown in Table 2.3.

Table 2.3 Properties of lime used in cold recycling

Parameter	Standard	Value
Designation	EN 549 [35]	CL 90-S
$Ca(OH)_2$ [%]	EN 459-2 [36]	>90
MgO	EN 459-2 [36]	<5
pH	–	>12

2.3.3 Mixture Gradation

Cold recycled materials, like granular materials or asphalt mixes, need to have a selected gradation to achieve optimized properties. When milling an old asphalt pavement, the resulting RA gradation changes depending on the equipment used, the speed of milling, the temperature, the type of materials involved and the thickness of the milled layer [37]. The milling does change the gradation of the original asphalt mix, so it is important to measure the gradation of the milled material. Also, since the fines present in the original mix are bounded with the bitumen in form of mastic, it is often required to add mineral filler in the mix. The product of milling can be screened selecting the maximum particle size and the retained materials in form of lumps can be further crushed and screened to be appropriately reused. However, if the old pavement is mechanically ripped up, asphalt slabs need crushing and screening processes.

CRM refers to different grading envelopes depending on in which layer the mixture has to be used. Generally, the grading envelopes, especially when CIR recycling is undertaken, are more permissive (large tolerance ranges) than those imposed for HMA. The main reason is not to hinder the application of the technology. Obviously, a proper combination of binders has to balance accidental weakness in particle distribution.

Examples of specifications for gradation for cold bitumen emulsion material is presented in Fig. 2.2. As shown, envelopes are wide and allow a big variation in the gradation except for the AASHTO envelope. It should be noted that for AASHTO MP31 [38], the gradation requirements are separated into coarse, medium and fine aggregates fractions. The loose mix are produced from in-place cores and are supposed to be selected to match the expected gradation produced from the recycler.

The gradation of CRM has a direct impact on the workability and compactability in fresh stage and mostly on performance under traffic. It has been shown that dense graded distribution results in workable mixes with limited air voids and an adequate stone on stone contact. To limit the air voids to the lowest level possible, the Fuller-Thompson (FT) approach can be followed. In this approach, the gradation is optimized starting with a given coarse aggregate size and the smaller aggregates, in terms of size and quantity, are selected to fill the voids between the coarse particles.

The air voids in a HMA or CRM are directly related to their workability and compactability. Workability is related to the ability of the mix to be laid and the compactability is related to the energy required to compact the mix. For CRM, both workability and compactability are strongly influenced by the water content, but also by the gradation [40]. The amount of filler seems to have the biggest impact where high filler content enhances workability, but it does decrease the compactability. It is known that lower air voids result in better mechanical properties for CRM. However, it is unclear if, for fixed air voids, the gradation does have an impact on the properties. It has been shown that gradation is more important that the bitumen emulsion or cement content to improve mechanical properties [17, 41]. Also, Raschia et al. [40] have shown that when similar mix design is used (same air voids content,

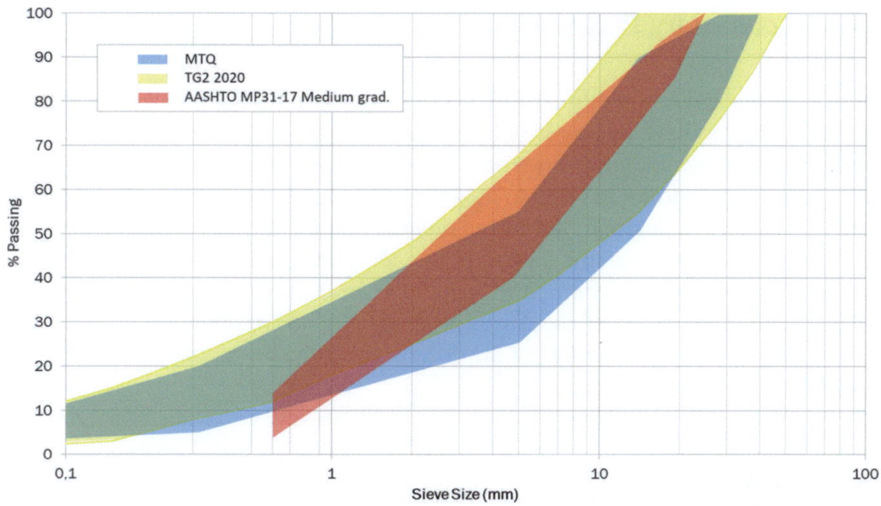

Fig. 2.2 Suitable gradation for emulsion treated materials according NQ 2560-114 [39], Sabita [13] and AASHTO MP31-17 [38]

same water, cement and residual binder), the FT gradation gives the highest indirect tension values compared to a dense-graded and gap graded mix.

2.3.4 *Selection of the Bitumen Addition Type*

As the name implies, CBTM contain cement and bitumen to give the material its cohesion. However, it is not specified if the bitumen is added as foam bitumen or as bitumen emulsion. Even if the mix design does change depending if foam or emulsion is used, their mechanical properties are often considered equal and no differences are taken into account when doing a pavement design [13]. However, some studies have shown differences in the properties of cured specimens [31, 42, 43].

The workability of foam or emulsion treated materials can be very similar, but their duration can differ. For foam bitumen, the workability can extend for hours in ideal conditions, where emulsion treated mixes are workable only before the breaking of the emulsion.

Another aspect to consider when comparing foam and emulsion treated CRM is how the bitumen bounds the particles together. With foamed bitumen, the coarse aggregates are spot welded together by the mastic that is composed of the fine on which the foamed bitumen is bounded with [23]. As for bitumen emulsion, the coverage of the coarse and fine aggregates is more complete.

2.3.5 Laboratory Procedures

For mix design and for quality control, CRM mixes have to be produced in the laboratory. In order to have laboratory mixes that are representative of field mixes, the mixing and compaction procedures have to be selected carefully. Also, CRM mixes have evolutive properties. In order to monitor the evolution of those properties and to match, as closely as possible, the curing process of the CRM also needs to be properly controlled. Unfortunately, there are no widespread agreements on the mixing, compacting or curing procedure. Information on the different laboratory procedures can be found in the different standards, in theses [44, 45] or other RILEM publications [9, 46]. Below is a summary of the differences in the laboratory procedures that can be found in those references.

2.3.5.1 Mixing

For the mixing part, what varies in different procedures are the mixing equipment used and the material preparation. For the duration, in most standards it is simply mentioned mixing until homogeneity, or less than 60 s [47]. For CRM, the mixing temperature does not need, by definition, to be as high as with HMA. Most standards mention room temperature or 25 °C for mixing. Depending on the standard, it is required to mix manually or with a mechanical mixer. In some cases, a pug-mill mixer is recommended in order to better simulate field conditions. There are little requirements regarding the batch size. However, there are specifications regarding the material preparation and the order in which the different components are mixed. There are basically two different approaches for CRM. In one case, the RA and the cement or lime are first mixed with water before adding the foam or the emulsion. This is the most common procedure. Not all the standards state if the RA needs to be dried before mixing. In the second method, part of the water is first mixed with the dry RA and then placed in a sealed container for 24 h to ensure that the RA absorbs water. Then, the saturated RA is mixed with cement or lime before adding foam or emulsion [15]. It should be noted that the authors are of the opinion that the latter approach is much more in line with the in-situ production of CRM.

2.3.5.2 Compaction

Laboratory compaction can't perfectly mimic the field compaction. This is not only because of the environmental conditions, but also because of a scale effect and because of the equipment used. Different laboratory equipment can be used to compact CRM. The most common equipments are the gyratory shear compactor (SGC), the Marshall hammer, the proctor rammer and the vibratory hammer. Slab compactors are also used for CRM, but mostly for research purposes. Also, with a given device, some differences can be specified. For example, with the SGC, it is possible to adjust angle of gyration to change the extent of shearing action during compaction. While some laboratories are continuing to use Marshall compaction, some specifications call for 75 blows on each side whereas other requires 50 blows on each side.

As previously mentioned, air voids do have a major impact on the properties and performances of CRM. This mean that compaction does influence the properties of the CRM and should be done properly.

If water is expelled from the mold during the compaction, this may be indicative of issues associated with compaction process, mix proportioning, including excess fluid content, and lack of homogeneity within the specimen. This phenomenon is discussed in detail in the volumetric part of this chapter.

2.3.5.3 Curing Procedures

A curing period is required with CRM that are treated with foam or emulsion. The curing duration and conditions do differ depending on the standard. The curing part is the most difficult to associate with field conditions since in the field, the cure depends on the environmental conditions, the thickness of the layer and the traffic. In the laboratory, since it is required to cure the specimens quickly for quality control, many standards ask for an accelerated cure in an oven. One to three days at 40, 50 or 60 °C is used to accelerate the cure that can take up to 14 days at room temperature. It should be noted that the cure is not necessarily completed after the specified period, but it is considered sufficient to quantify the properties of the CRM. Curing is often associated with the water evacuating the specimens, but the cure of emulsion treated CRM is not completed even when the water content becomes constant [48]. To accelerate or limit the rate of water loss and to simulate field conditions, it is possible to partially, or totally, seal the specimens during the cure so the water can only be evacuated from the top of the specimens [4].

2.4 Volumetric Approach Applied to CRM

2.4.1 Introduction

To ensure a good reproducibility and thermomechanical properties of CRM, an adequate mix design has to clearly specify the proportions of each constituent. Based on the volumetric approach applied for HMA, this section highlights the concepts of the volumetric approach applied to CRM.

Several elements are distinctive when comparing mix design of CRM to that of HMA, which differ depending on the use of bitumen emulsion or foamed bitumen, the presence of water and the mixture state (fresh or cured). Bitumen in form of emulsion has lubricating effects that allows RA to be adequately coated and CRM to be workable at ambient temperature. In case of foamed asphalt, the added water to the mix serve the role of providing lubrication and the fines in the mixture provides role of bitumen carrier for dispersion. When bitumen setting and mixture curing take place, mixture turns from the fresh state to the cohesive state.

The CRM cohesion increases over time depending on the evolution of the curing process. After compaction CRMs typically contain about 4–8% of water and the voids range between 10 and 15% [9, 49]. The rate of curing of CRM depends on several factors:

- External conditions such as temperature, moisture, drainage conditions, depth in the structure and sealing,
- Mixture composition such as water content, air voids, binders' type, dosage and ratio.

2.4.2 Assumptions

Considering the introduction of cement and water in the system, [15, 50] several initial assumptions are needed to extend the volumetric approach used for HMA mix design to that of CRMs. In particular, the concept of "Voids Filled with Liquids-VFL" is introduced, which is analogous to the concept of "Voids Filled with Bitumen-VFB" for HMAs.

The assumptions used for the volumetric approach applied to CRMs are:

1. Aggregates (RA and virgin aggregates) are considered in saturated surface dry condition for the lifetime of the product;
2. External pores of virgin aggregates could be assumed saturated of water and the surface of RA particle is covered by old bitumen, hence, no fraction of the added bitumen is absorbed by aggregates and RA;
3. RA particles are inert particles (black rock) therefore RA bitumen is considered to be a part of the aggregates and not considered as active binder;
4. The bulk volume of the CRMs after compaction (V^0_{mb}) is fixed over time;
5. Even if the bitumen coating on aggregate particles could be partial and water in the external pores of aggregates could potentially evaporate, aggregates always remain in an SSD state since a constant moist condition generally persists during service life. In this document, "dry" refers to CBTM that is free of water but whose aggregates are still in an SSD state.
6. The free water (or external water) represents the inter-granular water in the mixture. The free water consists of water added to improve compactability. When bitumen emulsion is used, the free water also includes the water brought into the mixture by the emulsion.
7. Absorbed water by aggregates is considered part of the aggregates and not free water.
8. The air voids content for CRMs includes the free water (inter-granular water).
9. During cement hydration part of water turns in cement hydration product.

Table 2.4 Main parameters to consider establishing the volumetric proportions of HMAs and CRMs

HMA	CBTM-EMU and CBTM-FOAM as evolving structure	
	Initial state (after compaction: t = 0)	Long-term state (after curing and some "α" degree of hydration)[a]
Basic parameters		
V_i (% of V_{mb}) VMA (% of V_{mb}) VFB (% of VMA)	V^{0}_{m} (% of V^{0}_{mb}) VMA^{0} (% of V^{0}_{mb}) VFL^{0} (% of VMA^{0})	V^{α}_{m} (% of V^{α}_{mb}) VMA^{α} (% of V^{α}_{mb}) VFL^{α} (% of VMA^{α})
Measured parameters on fresh mix		
G_{mm} (ρ_{mv}/ρ_{w}) [51] Weight of each component (aggregates: dry state)	Weight of each component (aggregates: saturated surface dried: SSD)	Weight of each component (aggregates: saturated surface dried: SSD)
Measured properties on compacted specimen		
G_{mb} (ρ_{bdry}/ρ_{w}) [52]	V^{0}_{mb}	V^{0}_{mb}
Calculated parameters based on a rational approach		
N/A	G^{0}_{mm} (ρ^{0}_{mv}/ρ_{w}) [51] (Maximum "dried" Specific gravity after compaction)	G^{α}_{mm} ($\rho^{\alpha}_{mv}/\rho_{w}$) [51] (Maximum "dried" Specific gravity after compaction, at "α" cement hydration)

[a] "α" represents the degree of the cement hydration: (no hydration: $\alpha = 0$; full cement hydration: $\alpha = 1$)

During laying down operations, water plays a key role in the CRM by acting as a lubricant but after compaction free water does not give any contribution to the mixture. After compaction, CRM experiences two different states: the initial state where free water is trapped in the microstructure and the long-term state, after the cure, when volumetric properties of a CRM are similar to those of an HMA. Table 2.4 illustrates the distinctive parameters that should be measured during HMA and CRM production to define their volumetric properties. The meaning of acronyms introduced in Table 2.4 are explained in detail in Sect. 2.4.3.

2.4.3 Initial State Analysis: Composition After Compaction

Figure 2.3 illustrates the volume composition of CRMs after compaction (initial phase) where the cement is considered as a filler.

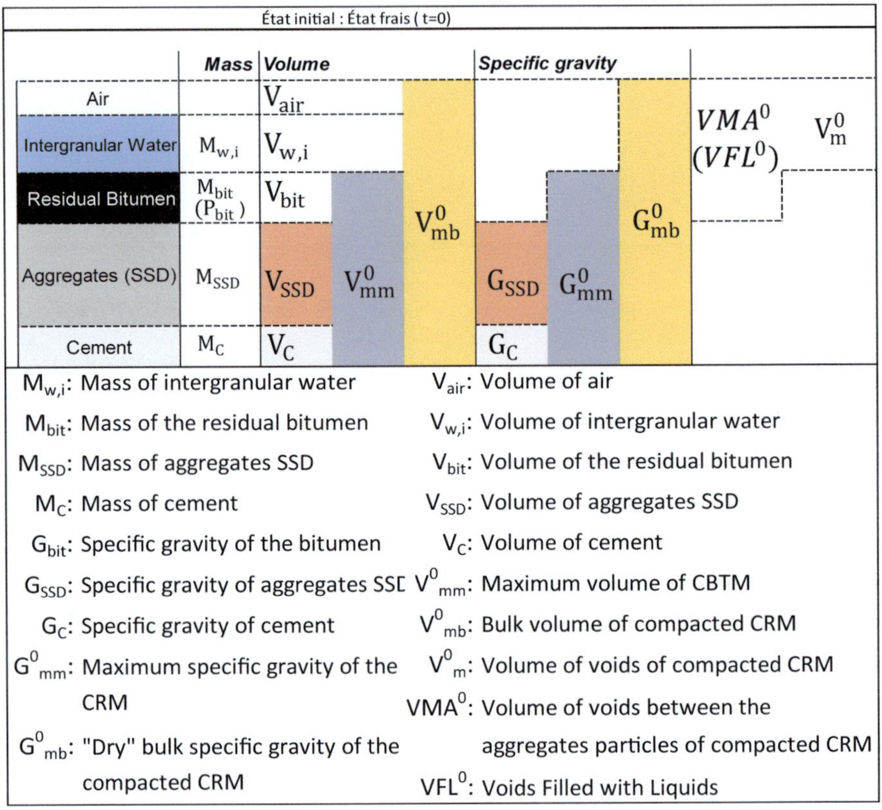

Fig. 2.3 Volumetric parameters for CRM: Initial state, "initial" state (t = 0)

In accordance to Fig. 2.3, the "dry" bulk specific gravity of compacted CRM can be written:

$$G^0_{mb} = \frac{\text{Dry weight of the compacted specimen}}{\text{Bulk volume of the compacted specimen}} = \frac{M_{SSD} + M_C + M_{bit}}{V^0_{mb}} \tag{2.1a}$$

$$G^0_{mb} = \frac{M_{SSD} + M_C + M_{bit}}{\frac{M_{SSD}}{G_{SSD}} + \frac{M_C}{G_C} + \frac{M_{bit}}{G_{bit}} + \frac{M_{W,i}}{G_W} + V_{air}} \tag{2.1b}$$

The "dry" bulk specific gravity of compacted CRM can be also calculated by the "wet" bulk specific gravity and the free water content (inter-granular water):

$$G^0_{mb} = \frac{\text{Wet weight of the compacted specimen}}{\text{Bulk volume of the compacted specimen} \cdot (1 + w)} \tag{2.1c}$$

The initial bulk volume (V^0_{mb}) is determined by geometric measurements of the specimen after compaction (Marshall, Proctor or SGC). For the "dry" CRM, the maximum specific gravity (G^0_{mm}) is then calculated by considering the mass of the constituents in the compacted specimen, namely:

$$G^0_{mm} = \frac{M_{SSD} + M_C + M_{bit}}{V^0_{mm}} = \frac{M_{SSD} + M_C + M_{bit}}{\frac{M_{SSD}}{G_{SSD}} + \frac{M_C}{G_C} + \frac{M_{bit}}{G_{bit}}} \tag{2.2}$$

Depending on the G^0_{mb} and G^0_{mm} values, the volume of voids (V^0_m):

$$V^0_m(\%) = \frac{G^0_{mm} - G^0_{mb}}{G^0_{mm}} \cdot 100 = \frac{V_{air} + V_{W,i}}{V_C + V_{SSS} + V_{bit} + V_{air} + V_{W,i}} \cdot 100 \tag{2.3}$$

The void in the mineral aggregate ($VMA^0(\%)$) for a CRM compacted sample, which is the void space between aggregate particles, includes the volumes of air, free water and emulsion or foamed bitumen (not including aged bitumen in RA) and is expressed as a percentage of the total volume of the specimen. It can be written:

$$VMA^0(\%) = V^0_m(\%) + \frac{G^0_{mb} \cdot P_{bit}(\%)}{G_{bit}} \tag{2.4}$$

where $P_{bit}(\%)$ is the emulsion or foamed bitumen mass expressed in accordance with the "dry" mass of the compacted CRM, i.e.:

$$P_{bit}(\%) = \left(\frac{M_{bit}}{M_{"dry" CRM}}\right) \cdot 100 = \left(\frac{M_{bit}}{M_{SSD} + M_C + M_{bit}}\right) \cdot 100 \tag{2.5}$$

The fraction of the granular skeleton voids filled with water and emulsion or foamed bitumen is then written:

$$VFL^0(\%) = \frac{V_{W,i} + V_{bit}}{V_{bit} + V_{air} + V_{W,i}} \cdot 100 = \frac{\frac{M_{W,i} \cdot G^0_{mb}}{G_W} + \frac{M_{bit} \cdot G^0_{mb}}{G_{bit}}}{VMA^0\%} \cdot 100 \tag{2.6}$$

V^0_m, VMA^0 and VFL^0 parameters are rational quantities determined after compaction. They will be used both to validate the formulation of CRM and for quality control under production.

2.4.4 *Long-Term Analysis*

During the curing process, the thermomechanical properties of CRM change as a part of the pore water escapes from the mixture and another part reacts with the

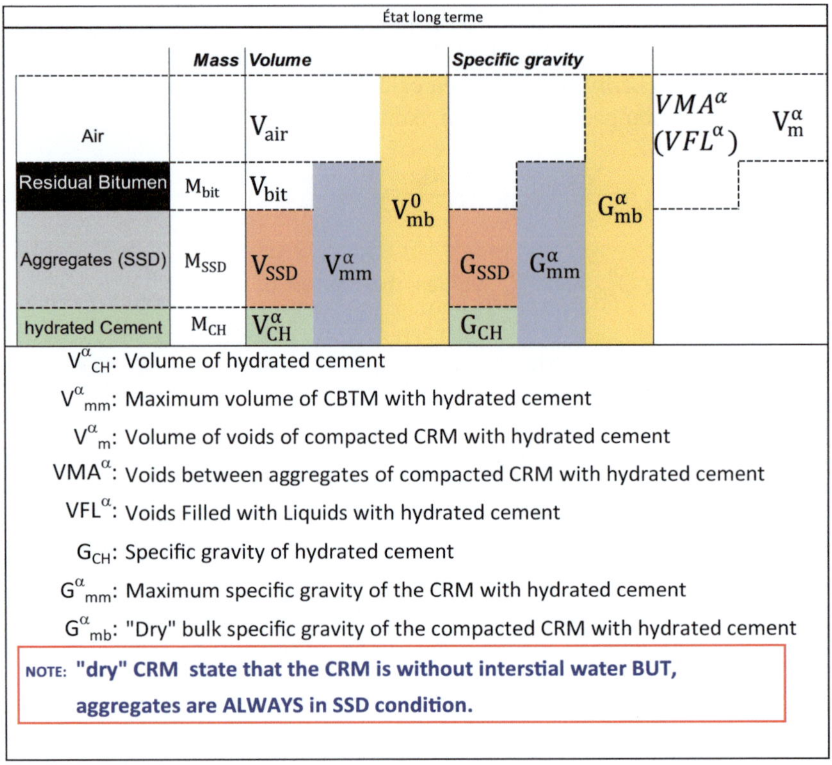

Fig. 2.4 Volumetric parameters for CRM: Long-term state (under service; $t = t$)

Portland cement. Accordingly, the absolute volumes of the constituents change over the long-term. Figure 2.4 shows long-term CRM volume composition.

As we have previously underlined, the "dry" bulk volume of the CRM after compaction (V^{0}_{mb}), remains constant over time (post-compaction considered negligible): only the volume of interstitial pore water will turn into air voids, due to evaporation process and cement hydration to generate a new solid phase: hydrates. In the field, the water content can change due to environmental conditions (precipitations, seepage or evaporation). In the long term, cement hydration progress and hydrates will take place in the microstructure (V^{α}_{CH}). The "α" parameter, introduced in Fig. 2.4 accounts the cement hydration level.

The main hydrate, calcium silicate hydrated (CSH), is a porous solid, like a gel type solid. According to cement chemistry concepts, for Portland-type cements the mass of hydrates (M^{α}_{CH}) is defined as:

$$M^{\alpha}_{CH} = M_{C}(1 + 0.42\alpha) \tag{2.7}$$

where M_C is the mass of cement and α is the degree of hydration.

The CSHs are composed of non-evaporable water (bounded water and water confined in the gel pores) and evaporable water contained in the capillary pores, which are created from the cement hydration process ("*Le Chatelier*"). The "*so-called capillary water*" in concrete chemistry within the hydrates is here considered as part of the free water of the CRM, therefore it is evaporable. Consequently, only the volume of hydrates (V_{CH}^{α}) is shown in Fig. 2.4, when the volume of the capillary pores being considered as a volume of air in the CRM. The volume of hydrates (V_{CH}^{α}) is then expressed as follows:

$$V_{CH}^{\alpha} = \frac{M_C}{G_C} \cdot (1 + 1.12\alpha) \tag{2.8}$$

Ultimately, the cement hydration has the effect of generating voids in the CRM since the hydrate products occupy a smaller volume than the absolute volume of the cement and the water involve in the chemical reaction. The "dry" bulk specific gravity of CRM following hydration (α) can be written as follows:

$$\begin{aligned} G_{mb}^{\alpha} &= \frac{M_{SSD} + M_{bit} + M_C(1 + 0.42\alpha)}{V_{mb}^0} \\ G_{mb}^{\alpha} &= \frac{M_{SSD} + M_{bit} + M_C(1 + 0.42\alpha)}{\frac{M_{SSD}}{G_{SSD}} + \frac{M_{bit}}{G_{bit}} + \frac{M_C}{G_C} + \frac{M_{W,i}}{G_W} + V_{air}} \end{aligned} \tag{2.9}$$

$$G_{mb}^{\alpha} = \frac{[M_{SSD} + M_{bit} + M_C(1 + 0.42\alpha)] \cdot \left[1 - \frac{V_m^0(\%)}{100}\right]}{\frac{M_{SSD}}{G_{SSD}} + \frac{M_{bit}}{G_{bit}} + \frac{M_C}{G_C}} \tag{2.9a}$$

$$\begin{aligned} G_{mm}^{\alpha} &= \frac{M_{SSD} + M_{bit} + M_C(1 + 0.42\alpha)}{V_{mm}^{\alpha}} \\ G_{mm}^{\alpha} &= \frac{M_{SSD} + M_{bit} + M_C(1 + 0.42\alpha)}{\frac{M_{SSD}}{G_{SSD}} + \frac{M_{bit}}{G_{bit}} + \frac{M_C}{G_C} \cdot (1 + 1.12\alpha)} \end{aligned} \tag{2.10}$$

Figure 2.5a shows the evolution of the maximum specific gravity (G_{mm}^{α}) and that the bulk specific gravity (G_{mb}^{α}) of the "dry" CBTM according to the cement hydration progress of a specific CBTM given in Fig. 2.5b. Figure 2.5a shows that the value of the maximum "dry" specific gravity of CBTM (G_{mm}^{α}) decreases as the cement hydration progress due to the increase in the volume of hydrates, while the "dry" bulk specific gravity (G_{mb}^{α}) increases due to the fixation of part of the free pore water in hydrates. Finally, the content of air voids in the CBTM decreases as the hydration of the cement develops.

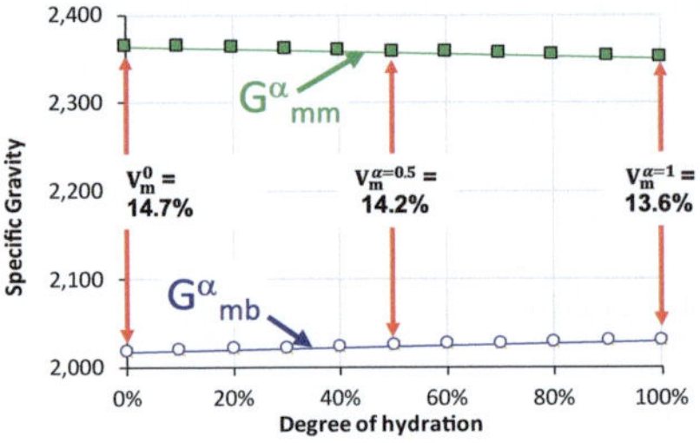

a)

Materials	Specific Gravity		Mass (g)
RA (NMS: 16mm)	SSD	2,440	963,1 M_{SSS}
Filler	SSD	2,650	60,9 M_{SSS}
Cement	Dry	3,150	15,2 M_C
Adding Bitumen[(1)]	n/a	1,015	30,4 M_{bit}
Integranular water[(2)]	n/a	1,000	30,4 $M_{w,i}$
LAB measurements on SGC compacted sample			
Sample Dia.	(cm)	10,00	
sample height at 100 (ng^0)	(cm)	6,75	
V^0_{mb}:	(cm^3)	534,1	
Calculated values d^0_{mb}:	n/a	2,018	eq. (1)
d^0_{mm}:	n/a	2,246	eq. (2)
V^0_{mm}:	(cm^3)	14,7	eq. (3)
VMA^0	(%)	20,3[(3)]	eq. (4)
VFL^0	(%)	56,1	eq. (6)
[(1)] Source: Bitumen emulsion at 60% solid (%mass)			
[(2)] Including emulsion water			
[(3)] avec P_{bit}(%) = 2,85% eq. (5)			

b)

Fig. 2.5 Example of application: **a** variation of the maximum and the bulk specific gravity of the "dry" CBTM-EMU in accordance to cement hydration, **b** example of CBTM-EMU formulation

2.4.5 *Mix Design*

To formulate a CRM from a given RA, firstly, its gradation is adjusted, if needed, according to local requirements using virgin aggregates, including filler. In Europe, the RA gradation is often corrected with virgin aggregates following the densest particle packing as defined by the Fuller/Thompson (FT) relationship. In a second step, the procedure compares several mixtures containing different water content and an expected cement content but without bituminous binder. The water content which allows the maximum dry density to be reached is defined as the optimal water content. In the third step, several mixtures with the optimal water content and different combinations of residual bitumen (2.0–3.0%) and cement (0.5–3.0%) are produced, cured, tested and selected to comply with the mechanical specifications (ITS, ITSM and others) [53].

Using the SGC, the volumetric approach can be easily applied. Of course, the bulk volume evolves under processing (V^{ng}_{mb}) and their tracking is done by continuously measuring the height of the specimen at each number of gyrations (ng). For SGC, the initial bulk volume (V^0_{mb}) is fixed according to a number of reference gyrations (ng^0). Although the ng^0 value is currently arbitrary, several specifications estimate that the compaction energy on site is similar to that reached after 100 gyrations under SGC compaction [54]). The validation of this last point is required.

In accordance with the previous equations, V^{ng}_m, VMA^{ng} and VFL^{ng} values can be expressed as a function of the compaction energy (ng). Figure 2.6 shows an example of the evolution of V^{ng}_m, VMA^{ng} and VFL^{ng} parameters of a CBTM-EMU under SGC. The values of V^0_m, VMA^0 and VFL^0 are noted at ng^0. It is important to remark that

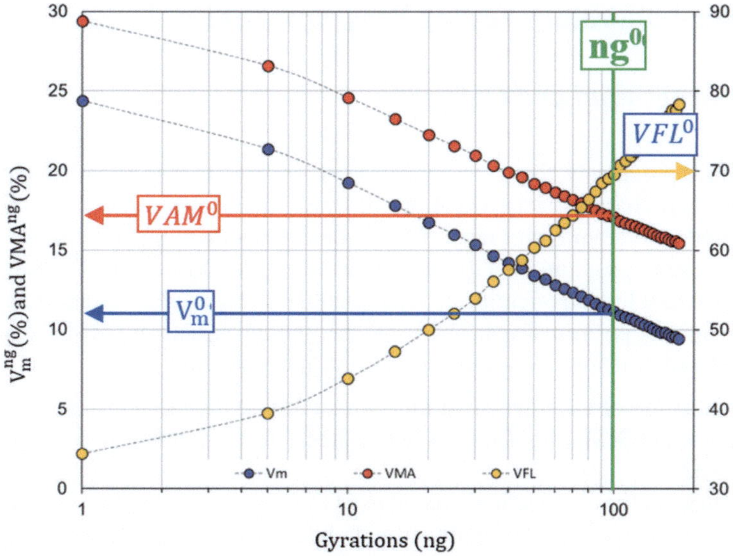

Fig. 2.6 Example of the evolution of V_m^{ng}, VMA^{ng} and VFL^{ng} parameters of a CBTM-EMU under SGC compaction. V_m^0, VMA^0 and VFL^0 at $ng^0 = 100$ gyrations are illustrated

when VFL exceeds 85–90% water is expelled from the mold. Since often bitumen droplets and fines are expelled as well, this leads to a change in the CRM composition. Therefore, is important to keep the VFL below the aforementioned limits [55, 56].

2.4.6 *Quality Control During Production*

From the moment the CRM formulation has been validated in the laboratory (V_m^0 and ITS/ITSM mechanical performance), the quality control in production will consist in verifying the plant production records and in measuring in-situ the “dry” bulk specific gravity of the CRM (G_{mb}^0). To do this, one can refer to the measurement of the density according to the standard ASTM D1556/M1556-15 [57]. In addition, ad hoc plant measurements of G_{mb}^0 with SGC test allow rigorous monitoring of CRM production.

2.5 Inter-Laboratory Testing Program (ILTP)

2.5.1 Background and Objectives

As mentioned previously, there are several aspects related to cold recycled materials mix design and characterization that are not well understood. Based on information gathered in literature and from international participant institutions, an inter-laboratory testing program (ILTP) was launched in the TG1 of the TC-RAP. The objective was to achieve a scientific understanding of cold recycled materials, in order to ensure superior performances. The ILTP focused on aspects of CRM production that could be of practical and industrial interest, more exactly:

a. The evaluation of the fragmentation test as a tool to highlight differences in the properties of CRM with different RA sources. At this regard, two RA sources were used;
b. The impact of the compaction method used for laboratory specimens preparation (Marshall hammer, proctor hammer and shear gyratory compaction using two diameters);
c. The influence of different types of laboratory curing (free surface drying-FSD; partially surface drying-PSD);
d. The effect of the type of cold binder (foamed bitumen and bitumen emulsion);
e. Implement volumetric approach for mix design and quality control.

To collect results as homogenous as possible from the participating laboratories, and to focus strictly on the objective mentioned above, the CBTM materials were produced using the same raw materials. The details of the ILTP are presented in the next sections.

2.5.2 Methodology

The ILTP aims to reach multiple objectives involving a high number of variables. In order to obtain comparable results from several laboratories and operators, some parts of the experimental plan and the methodology were fixed in guidelines spread to the participants. In addition, each laboratory had the possibility to develop the complete experimental plan or to study only particular aspects considered more interesting (for example: one RA source, two compaction methods, bitumen emulsion, one type of curing and so on). In this way it was also possible to highlight the interest of the international participants on specific aspects of CRM materials. The general points characterizing the methodology are the following:

- The raw materials (RA, filler, cement, bitumen emulsion and foamed bitumen) were collected in one laboratory and shipped to the laboratories-participants;
- A common procedure for specimens mixing, compaction and curing, was prepared. The mixing protocol followed starts with the blending of RA and filler

and the addition of absorption water, to reach the SSD condition. After a period of around 24 h, cement, water and binder (bitumen emulsion or foamed bitumen) were added in this order. At each addition, the mixture was mixed for at least 60 s. Cement content was fixed at 1.5% by solids mass (RA, filler and cement). Bitumen content (emulsion residue or foamed bitumen) was fixed at 3.0% by solids mass too;

- The two RA sources were chosen according to fragmentation test results. The two materials have different gradation, but a correction with limestone filler was done to follow the maximum density curve;
- The compaction methods chosen are the most common for laboratory compaction of CRM: Marshall hammer, Proctor rammer, Shear Gyratory Compactor (SGC)—Ø100 mm (SGC10) and Ø150 mm (SGC15). A specific mix design, involving volumetric approach, was developed to fix water contents and compaction energies to reach similar values of air voids among the four compaction methods. Further details are given below;
- Two curing protocols were investigated: a free surface drying (FSD) curing and a partially sealed drying (PSD) curing. In the first case (FSD) the water evaporation occurs from all the surfaces of the specimen, whereas in the second case (PSD) the round surface of the specimen is sealed with plastic foil. This procedure, more representative of field condition, allows water evaporation only from the top flat surface of the specimen.

In both cases, the curing occurred for 14 days at 40 °C and, if it was possible, at controlled relative humidity of 55 $\pm$ 5%. During curing, water loss was monitored by weighing the specimens at 1, 3, 7 and 14 days;

- Mechanical tests were performed on cured specimens. In particular Indirect Tensile Strength (ITS-EN 12697-23 [58]/ASTM D6931 [59]), Indirect Tensile Strength Ratio (ITSR-EN 12697-12 [60]/ASTM D4867 [61]), Indirect Tensile Stiffness Modulus (ITSM-EN 12697-26 Annex C [62]). ITSM tests was carried out at 3 temperatures: 2, 10 and 20 °C.

2.5.3 Participating Labs

Ten countries were involved in the ILTP: University of San Marino (San Marino); University of New Hampshire (USA); University of Padua (Italy); University of Antwerp (Belgium); Vienna University of Technology (Austria); University of Parma (Italy); Braunschweig University of Technology (Germany); University of Nottingham (UK); University of Sao Paulo (Brazil); Université Gustave Eiffel (France); École de technologie supérieure (Canada). The ILTP was a big opportunity to exchange with 12 laboratories from 10 different countries on the problematics linked to the laboratory production of CBTM and to share a common procedure to enhance reproducibility. The tests performed in each laboratory differ according to the equipment available.

2.5.4 *Raw Materials*

The raw materials used in every lab of the ILTP (RA, cement, filler and bitumen) are the same materials provided by the LCMB at ÉTS. Only the water used in the mixes was not provided. Also, as mentioned before, the CBTM mix design is the same for all the lab and it is provided by the LCMB at ÉTS, and the mix design is adapted according to the compaction method (Marshall, Proctor or SGC).

2.5.4.1 Reclaimed Asphalt: RA

One of the objectives of the ILTP was to evaluate the efficiency of the fragmentation test for the classification of different RA sources, and consequently to investigate the influence of RA source on CBTM mechanical properties. The fragmentation test protocol was described in the RILEM TC237-SIB: *Recommendation of RILEM TC237-SIB on Fragmentation Test for Recycled Asphalt* [18].

According to this, several RA sources were initially considered to select the more appropriate to be used to produce CBTM mixes. The testing protocol consists in following the Modified Proctor procedure and measuring the percentage of fines produced. This amount corresponds to the passing to a control sieve (1.7 mm) and it is indicated as PCS (passing through a control sieve) index, which is used to discriminate different RA sources. PCS indexes results for the RA sources investigated are reported in Table 2.5 and Fig. 2.7.

The results in Fig. 2.7 show significant differences depending on the source of RAs tested. Note that the RILEM recommendation is to evaluate fragmentation resistance at 5 °C. However, some RAs were tested at 20 C. It has been observed that PCS values at 20 °C are always lower than those obtained at 5 °C [18] which allows a conservative classification from the results at 20 °C.

In order to test significantly different RAs in terms of fragmentation test results, RA1 from Cooperativa Braccianti Riminese-CBR in Italy and RA2, from East Alabama Paving-EAP in the USA were selected. RA1 was sampled from a stockpile formed by milled material coming from a high traffic road, which was built with the highest aggregate quality possible (especially in terms of fragmentation and polishing). On the other hand, RA2 should come from a base course, which is usually made with aggregates of lower quality. For the purpose of the ILTP, 2.5 tons of RA1

Table 2.5 RAP sources classification (PCS index)

	Categories			
	B	C	D	E
$PCS_{5/10}$ (%) @ 20 °C	≤7	≤9	≤11	≤14
RA sources investigated	CBR (IT)	Chambly (QC) Bauval (QC) St-Isidore (QC)	St Bruno (QC) Sherbrooke (QC)	NCAT (US)

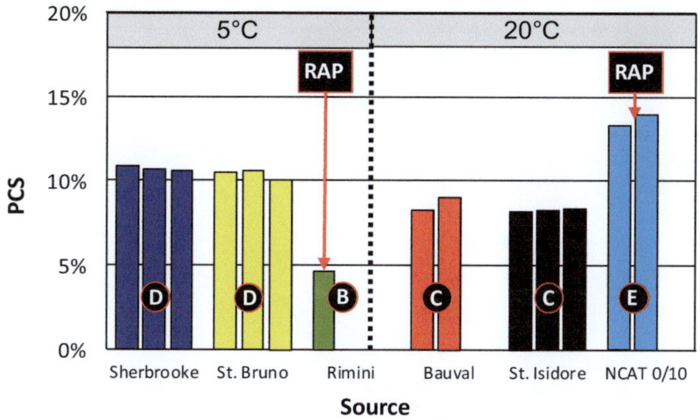

Fig. 2.7 Fragmentation test results (PCS%) obtained on fraction 5/10 of different sources of RA

and 1.0 ton of RA2 were collected and sampled at LCMB-ÉTS. Table 2.6 summarizes the main properties of both RAs. Black curves and white curves of the two RA used are reported in Fig. 2.8.

Reclaimed asphalt samples were shipped in humid state in plastic buckets of 20–25 kg to each participating laboratory. In order to minimize uncertainties related to the RA gradation into the CBTM mixes produced in different labs, each bag was quartered [68, 69] to reduce gross RA sample to each test sample size and stored in a seal plastic bags till mixing. A minimum of one particle size analysis per received

Table 2.6 Properties of RA sources used

Property	Standard	Unit	RA1 (i = 1)	RA2 (i = 2)
Binder content	ASTM D6307 [63]	%	5.51	5.43
Nominal maximum particle dimension	ASTM D448-03 [64]	mm	16	10
Maximum specific gravity (G_{mm})	ASTM C127-128 [65]	–	2.486	2.498
SSD bulk specific gravity (G_{SSD}^{RAP})	EN 1097-6 [66]	–	2.389	2.443
Dry bulk specific gravity (G_{mb}^{RA})[a]		–	2.362	2.417
Water absorption: abs	ASTM C127-128 [65]	%	1.10	1.08
Fragmentation @ 5 °C (5/10)—PCS	ASTM D1557 [67]	%	7.6	–
Fragmentation @ 20 °C (5/10)—PCS	ASTM D1557 [67]	%	6.7	13.9

[a] $G_{mb}^{RAi} = \frac{G_{SSD}^{RAi}}{(1+abs\%/100)}$

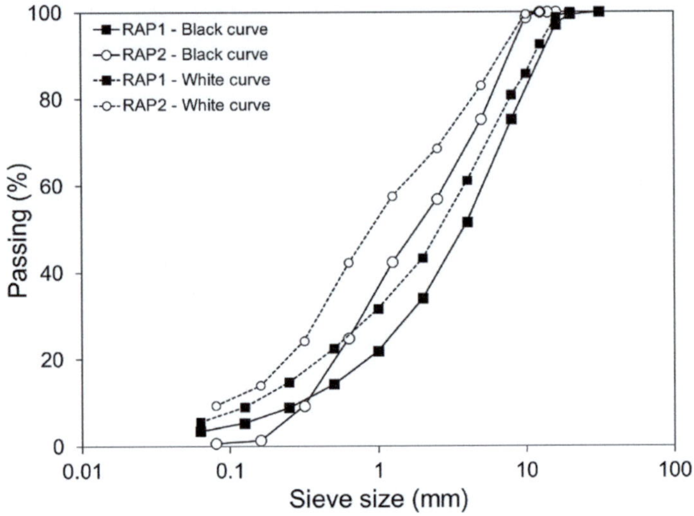

Fig. 2.8 Black curve and white curve of RA1 and RA2

bag was performed to validate RA gradation (RA1 or RA2) in accordance with the following standard envelope (Fig. 2.9). If necessary, test sample bags were corrected prior to mix.

Standard envelopes were determined based on the gradation limit values established statistically from all shipped RA samples that are supplied to all participant labs with a 99.9% confidence interval.

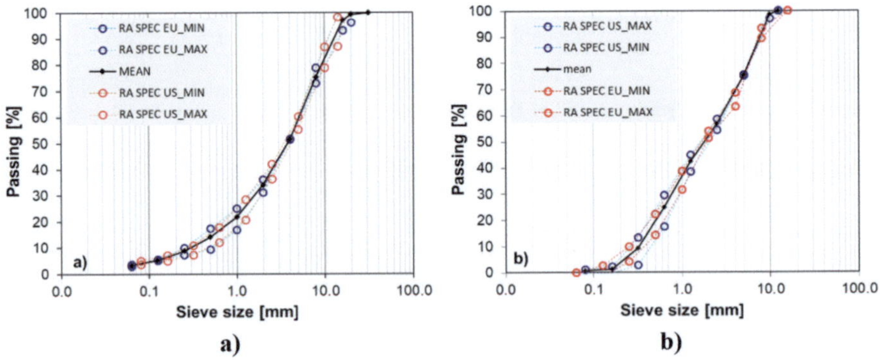

Fig. 2.9 Average gradation for the received RA material with the envelope of tolerance: **a** RA1, **b** RA2

Table 2.7 Bitumen emulsion properties

Emulsion properties	Standard	Unit	Value
Particle polarity	EN 1430 [71]	–	+
Breaking value	EN 13075-1 [72]	–	>170
Cement stability	EN 12848 [73]	g	≤2
Binder content (by water content)	EN 1428 [74]	%	58–62
Binder content (by distillation)	EN 1431 [75]	%	≥58
Distilled oil content	EN 1431 [75]	%	≤2
Efflux time (2 mm @ 40 °C)	EN 12846 [76, 77]	s	15–70
Residue on sieving (0.5 mm)	EN 1429 [78]	%	≤0.2
Settling tendency (after 7 days)	EN 12847 [79]	%	≤10
Adhesivity	EN 13614 [80]	%	≥75
Residual bitumen properties			
Needle penetration @ 25 °C	EN 1426 [81]	0.1 mm	
Oftening point	EN 1427 [82]	°C	≥43

Source Iterchimica Srl—Italy

2.5.4.2 Bitumen Emulsion

The bitumen emulsion used in this work is a cationic slow setting type [70] provided by Iterchimica Srl (Italy). The main properties are listed in Table 2.7. In the framework of the ILTP, as the emulsion could break during shipping, an international producer was the best choice to manage the shipping from the various worldwide facilities. It is important to remark that the bitumen contained in the emulsion is not the same used to produce foamed bitumen.

2.5.4.3 Foam Bitumen

Bitumen used to produce foam in this study has 64 mm × 10^{-1} value of penetration [81] and a softening point of 52 °C [82]. It has been provided by Nynas AB, which take care of the shipping to the participating labs. Preliminary tests had a fixed expansion ratio of 10 and 6 s as half-life at 170 °C (water content of 1.9% by bitumen weight). The water used to produce foamed bitumen is considered negligible in the total amount of water in the CRM material.

2.5.4.4 Filler

The filler used to correct RA gradation is a limestone filler obtained by a production plant dust collector. Table 2.8 gives physical properties of the filler used by the

Table 2.8 Limestone filler properties

Filler properties	Standard	Unit	Value
Passing at 80 μm sieve	ASTM D448-03 [64]	%	90.9
Rigden voids	EN 1097-4 [28]	%	50.1
Density ρ_f	EN 1097-7 [25]	g/cm^3	2.650
Stiffening power $\Delta_{R\&B}$	EN 13179-1 [29]	°C	11.5

participating labs. The water absorption of filler was not measured and not considered in the CRM volumetric properties since it is considered negligible.

2.5.4.5 Portland Cement

The cement employed is a type GU [83] with compressive strength at 28 days of 43.9 MPa [84]. Cement properties are shown in Table 2.9.

2.5.5 Selection of CBTM Recipe: Mix-Design

The mix-design of CBTM mixtures for the ILTP consisted in three steps: the correction of the RA sources gradation, the determination of cement and bitumen content and the study of the optimum water content.

The target gradation of the aggregate blend to be used refer to the maximum density curves, which relate to different aggregates maximum size: for RA1, the maximum size diameter is 16 mm, while for RA2 is 10 mm. The maximum density curve is calculated based on the Fuller-Thompson approach:

$$P(\%) = 100 \cdot \left(\frac{d}{D_{max}}\right)^{0.45} \quad (2.11)$$

Table 2.9 Cement properties

Property	Standard	Unit	Value
Name	–	–	General use (GU)
Density	ASTM C188 [85]	g/cm^3	3.15
Blaine fineness	ASTM C204 [86]	cm^2/g	4100
Cumulative < 45 μm	ASTM C430 [87]	%	95.5
Initial setting time	CSA A3004-B2 [88]	Min	180
Final setting time	CSA A3004-B2 [88]	Min	277
Compressive strength at 28 days	CSA A3004-C2 [89]	MPa	43.9

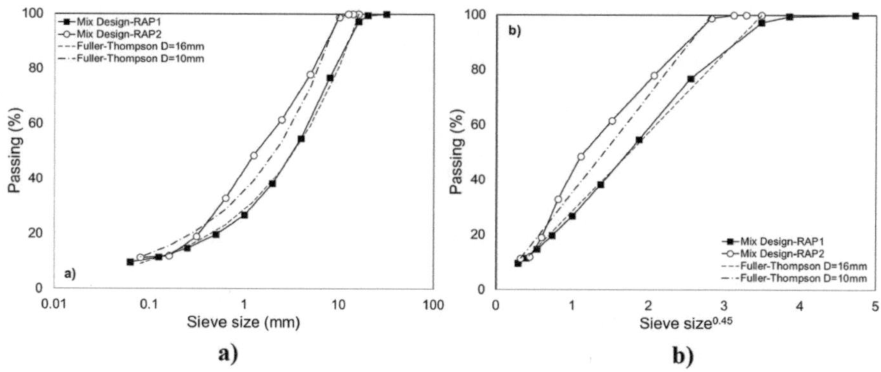

Fig. 2.10 Target gradation of the final aggregate blend: **a** logarithmic representation; **b** linear representation

where, D_{max} represents the size of the sieve where 100% of the particles are passing.

To meet this requirement, a certain amount of limestone filler shall be used. In particular, 5.5% and 10.0% of filler were used for RA1 and RA2, respectively. At this stage, cement is considered as a filler-sized particle aggregate, hence the gradations were adjusted considering a cement content of 1.5% for both RAs, by solids mass (RA, filler and cement). Ultimately, the aggregate blends of the CBTM materials were the following:

- CBTM-RA1: 93% of RA1, 5.5% of limestone filler and 1.5% of cement;
- CBTM-RA2: 88.5% of RA2, 10.0% of limestone filler and 1.5% of cement.

Gradations of the final aggregate blends are shown in Fig. 2.10.

The cement content was fixed at 1.5% by solids mass to respect the limit presented in Fig. 2.10.

Two (2) basic formulations were considered in the ILTP: (a) RAi_CBTM-EMU, when bitumen emulsion is used and, (b) RAi_CBTM-FB, when foamed bitumen is used (*i: to identify the RA source. i = 1 or 2*). To have comparable results between the two formulations, the same amount of bitumen (or residual bitumen from the emulsion) was used for all cases. The residual amount of bitumen was fixed at 3% by weight of the dry solids mass (RA, filler and cement) for all mixes. This amount was fixed to keep a bitumen/cement ratio around 2. In addition, it was assessed that such amount of bitumen did not influence significantly volumetric properties of CBTM mixtures produced with the two RAs. To do that, the compactability of such mixtures were studied by using SGC (100 and 150 mm diameters), employing both bitumen emulsion and foamed bitumen, and fixing the total water content at 3% by solids mass. The compaction was done at 180 gyrations (ng), 600 kPa of pressure, 30 rpm, and the voids in the mixture V_m were evaluated according to Eq. 2.3. The results obtained are collected in Table 2.10.

It is noted that the mean value of V_m is 9.3% with a standard deviation of 1.0% which is quite acceptable range to consider similar volumetric properties among the

Table 2.10 Compactability study for bitumen content

RAP source	Binder type	SGC diameter (mm)	V_m (%) at 180 ng
RA1	Bitumen emulsion	100	9.0
		150	N/A
	Foamed bitumen	100	10.8
		150	8.5
RA2	Bitumen emulsion	100	9.4
		150	N/A
	Foamed bitumen	100	9.5
		150	8.2

different mixtures. In addition, such value of voids in the mixture (with no material loss during compaction) is considered low with regards to CRM materials. This could be a good indicator of a well-designed aggregate blend, and a good starting point to calibrate the other compaction methods used: Marshall hammer and Modified Proctor.

In order to obtain similar voids content between the three compaction methods, the optimum water content was established for both Marshall hammer and Modified Proctor, following standards procedures. For simplicity, this step was performed with only bitumen emulsion, and it was assumed that the results could be applied also to mixtures produced with foamed bitumen. The optimum water content with Marshall hammer was defined according to LC 26-002 [90], by assuming a bitumen content of 3% by solids mass. According to this procedure, 5 water contents were tested (4, 5, 6, 7, 8%), and the maximum dry bulk density reached with 50 blows each side determined the optimum water content.

A similar approach was followed with the Modified Proctor procedure, in which 5 water contents were tested (3, 4, 5, 6, 7%) and the maximum dry bulk density reached with 56 blows per layer on 5 layers determined the optimum water content.

The same procedure to determine similar volumetric properties was followed for both RA1 and RA2. The results are shown in Figs. 2.11 and 2.12.

For RA1, the target range of dry bulk density was from 2.011 to 2.058 g/cm^3, which means 14.3 $\pm$ 1% of air voids considering a maximum specific gravity of 2.374. Based on Fig. 2.11, it is possible to reach a dry bulk density between this range with 4% of total water for Marshall hammer, 4% of total water for Modified Proctor rammer and from 45 to 80 gyrations for SGC. It is possible to compact with such number of gyrations by fixing the specimen height at 67.7 (samples of 100 mm diameter) and 76.7 (samples of 150 mm diameter).

For RA2, the target range of dry bulk density was from 2.023 to 2.070 g/cm^3, which means 14.3 $\pm$ 1% of air voids considering a maximum specific gravity of 2.388. Based on Fig. 2.12, it is possible to reach a dry bulk density between this range with 4.5% of total water for Marshall hammer, 4.0% of total water for Modified Proctor rammer and from 25 to 50 gyrations for SGC. It is possible to compact with such number

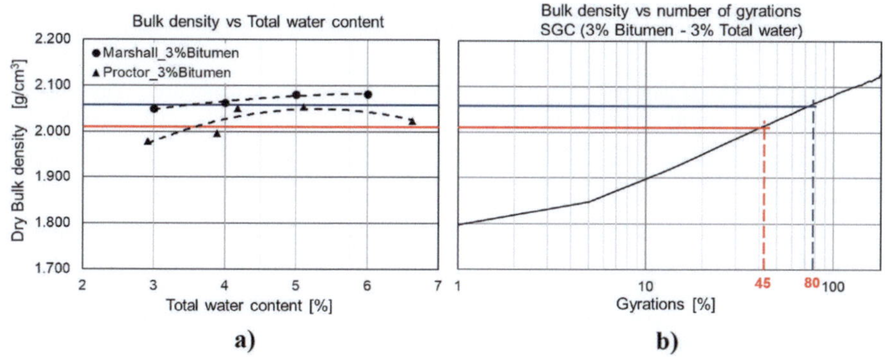

Fig. 2.11 Relationships between dry bulk densities of the compacted specimens for RA1 (Marshall, Modified Proctor and Gyratory Compactor)

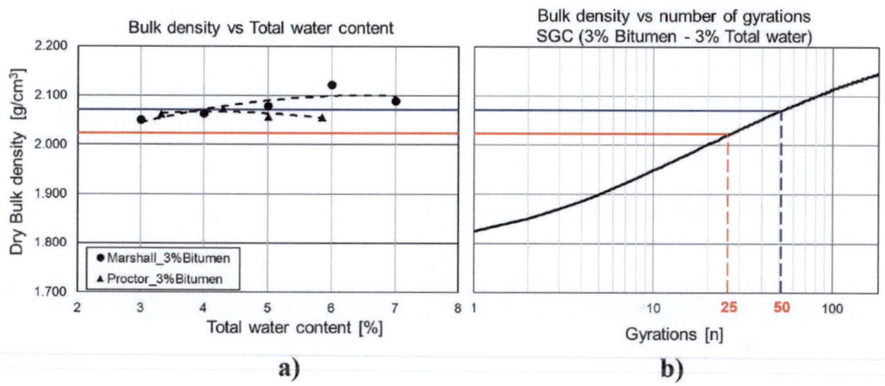

Fig. 2.12 Relationships between dry bulk densities of the compacted specimens for RA2 (Marshall, Modified Proctor and Gyratory Compactor)

of gyrations by fixing the specimen height at 67.1 (specimens of 100 mm diameter) and 75.9 (specimens of 150 mm diameter).

In conclusion, an overall on the compaction protocol and the target volumetric properties is given in Table 2.11.

Based on the protocol just mentioned, target CBTM-EMU and CBTM-FB formulations were fixed and provided to each laboratory as summarized in Table 2.12. The target voids in the mixture are 14.3 ± 1%.

Table 2.11 CBTM mixes compaction protocol with target volumetric properties

RA1			RA2		
Air voids [%]	Maximum dry density [g/cm³]	Dry Bulk density [g/cm³]	Air voids [%]	Maximum dry density [g/cm³]	Dry bulk density [g/cm³]
15.3	2.374	2.011	11.9	2.388	2.023
13.3		2.058	13.9		2.070
Compaction protocol			Compaction protocol		
Marshall	3.0% Bitumen 4.0% Total water 50 blows each side		*Marshall*	3.0% Bitumen 4.5% Total water 50 blows each side	
Proctor	3.0% Bitumen 4.0% Total water 5 layers 56 blows each layer		*Proctor*	3.0% Bitumen 4.0% Total water 5 layers 56 blows each layer	
SGC	3% Bitumen 3% Total water Specimens' height 67.7 (100 mm diameter) 76.7 (150 mm diameter)		*SGC*	3% Bitumen 3% Total water Specimens' height 67.1 (100 mm diameter) 75.9 (150 mm diameter)	

2.5.6 *Inter-Laboratory Testing Protocol (ILTP)*

The laboratory experimental work was described and summarized in guidelines that were shared with the participants before the shipping of raw materials. In these guidelines mixing, compaction and testing protocols were explained.

2.5.6.1 Mixing Protocol

In order to have mixtures as homogenous as possible, mechanical mixing was always preferred than manual mixing. However, if mixtures of less than 1000 g needed to be produced, manual mixing was allowed.

Bitumen Emulsion

RA material was initially oven dried at around 40 C. Subsequently, filler and absorption water were added, mixed and sealed in a plastic bag for at least 12 h. In a second step, cement, additional water and bitumen emulsion were added in this order, and mixing was performed at each addition for 30–60 s. For CBTM mixtures produced with bitumen emulsion, the batch mass was not fixed. The batch obtained was kept sealed in a plastic bag to slow down as much as possible the water evaporation, while the compaction process was taking place.

Table 2.12 Target CBTM-EMU and CBTM-FB formulations

		RA1					RA2				
			CBTM-EMU		CBTM-FB			CBTM-EMU		CBTM-FB	
			Batch of 3,000 g M_T (g)					Batch of 3,000 g M_T (g)			
		OD[3] (%)	ω% 3.0	ω% 4.0	ω% 3.0	ω% 4.0	OD[c] (%)	ω% 3.0	ω% 4.5	ω% 3.0	ω% 4.5
RA	M_{RAi}	93.0	2790.0				88.5	2655.0			
Mineral filler	M_{FILL}	5.5	165.0				10.0	300.0			
Cement-GU	M_C	1.5	45.0				1.5	45.0			
Absorbed water[a]	$M_{W,A}$	1.0	27.9				1.0	26.6			
Residual bitumen[b]	$M_{B,R}$	3.0	90.0				3.0	90.0			
Emulsion[d]	M_E	5.0	150.0		0.0		5.0	150.0		0.0	
Emulsion water[e]	$M_{W,E}$	2.0	60.0		0.0		2.0	60.0		0.0	
TOTAL Water[g]	$M_{W,TOT}$	–	90.0	1120.0	90.0	120.0	–	90.0	135.0	90.0	135.0
Mixing water[f,h]	$M_{W,M}$	–	2.1	332.1	62.1	92.1	–	3.4	48.4	63.4	108.4

[a] Water absorbed by the RA
[b] 3% of residual bitumen expressed in accordance to the total mass of dry solid component in the mix
[c] Oven Dried: 105 °C for VAg and 40 °C for RA
[d] For an emulsion with 60% of residual bitumen ($M_{B,R}$): 100 g of emulsion (M_E) ⇒ 60 g of bitumen ($M_{B,R}$)
[e] For an emulsion with 60% of residual bitumen ($M_{B,R}$): 100 g of emulsion (M_E) ⇒ 40 g of water ($M_{W,E}$)
[f] CBTM-EMU: The mixing water $M_{W,M}$ is obtained by the equation $M_{W,M} = M_{W,TOT} - M_{W,A} - M_{W,E}$
[g] The total water in CBTM-FB mixes includes the absorption water ($M_{W,A}$) and the mixing water ($M_{W,M}$). The foaming water is considered negligible
[h] CBTM-FB: The mixing water $M_{W,M}$ is obtained by the equation $M_{W,M} = M_{W,TOT} - M_{W,A}$

Foamed Bitumen

In case of CBTM mixtures with foamed bitumen, the batch size recommended was around 30 kg to obtain a homogenous blend in the pugmill mixer of the foaming unit. RA material was initially oven dried at around 40 °C, and subsequently mixed in the pugmill with filler and cement. In a second step, the total water was added manually, and the mixture was blended for 30–60 s. At this moment, the foaming process was started to spray the bitumen in the mixture and an additional mixing time of 30–60 s was allowed. The batch obtained was kept sealed in a plastic bag to slow down as much as possible the water evaporation, while the compaction process was taking place.

2.5.6.2 Compaction

Shear Gyratory Compactor (SGC)

The SGC protocol consists in two mould diameters (100 and 150 mm), pressure of 600 kPa, a speed of rotation of 30 rpm and a compaction angle of 1.25. As previously explained, the compaction process with SGC should be performed at a fixed specimen's height (or volume). The exact amount of CBTM material was poured inside the mould, and with the help of a suitable tool, an approximate plan surface was conferred to the loose material. After compaction is performed, the specimens are sufficiently resistant to be immediately extruded and placed on a support (plastic or aluminum plate) to record the weight loss along curing without manipulating the specimen.

Marshall Hammer

Marshall Hammer compactor was performed with 100 mm diameter moulds. Once the base, mould and collar were assembled, 1100 g of CBTM material were poured inside. With a suitable tool (spatula, steel stick, etc.) the material was pre-compacted with 15 hits along the mould diameter and 10 hits in the center. Compaction was performed with 50 blows on each side, after which the specimen should be kept in the mould at room temperature for about 24 h. After this period, specimens can be demoulded and placed on a support (plastic or aluminum plate) to record the weight loss along curing without manipulating the specimen.

Proctor Rammer

Modified Proctor rammer, generally used to investigate soils compaction, uses moulds diameter of 150 mm. Once the base, the mould and the collar are assembled, the material was poured to obtain 1 of 5 equal layers (around 1000 g). Compaction

Fig. 2.13 Proctor rammer specimen production

was performed with 56 blows (distributed on the specimen's surface) on each layer (Fig. 2.13). After compaction, the collar was removed, and all the exceeding material was scratched. At this point, the specimen should be kept in the mould at room temperature for about 24 h. After this period, specimens can be demoulded and placed on a support (plastic or aluminum plate) to record the weight loss along curing without manipulating the specimen.

Differences between the required dimensions for the tests and Proctor specimens required cutting and trimming for the Proctor specimens. The height of the proctor specimens is considered 120 mm. To conform ITS and ITSM test standards, Proctor specimens needed to get cut in half to have the least acceptable dimension to be comparable with other SGC 150 mm specimens.

2.5.6.3 Curing Process

The main research objective of this study is to evaluate a curing method which is an endeavor to simulate the field condition more accurately. As it is already stated, the main problem is to imitate the existing condition in the field, which is believed, doesn't let the water move out of the specimen lateral faces in curing period: water leaves out of the specimen only through the top and bottom sides of the specimens.

To evaluate this hypothesis, specimens were cured following two methods: a Free Surface Drying (FSD) curing and a Partially Sealed Drying (PSD) curing. The curing process took place in climatic chambers at 40 ± 2 °C and 55 ± 5% humidity for 14 days.

Free Surface Drying (FSD)

In Free Surface Drying (FSD), or the conventional existing curing protocol, which is used to practice accelerated curing process in the lab, after compaction and

Fig. 2.14 FSD curing

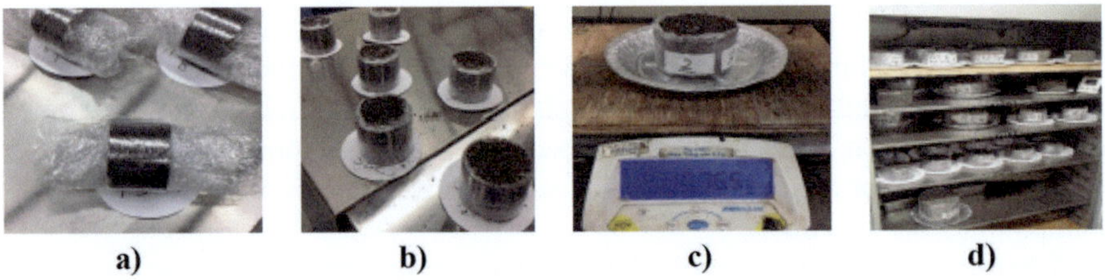

Fig. 2.15 PSD curing: **a** samples tight wrapped in tint, **b** sample upper surface is open, **c** samples are weighted, **d** the storage of samples

demoulding, the specimens are introduced to the direct air and humidity condition in the chamber without any boundaries (Fig. 2.14).

Partially-Sealed Surface Drying (PSD)

In the new protocol that is proposed in this research, it is tried to cover the lateral side of the specimens to prevent water from leaving from this part of the specimens. To do so, after mixing, compacting and demoulding the specimens, they are covered by 5–6 layers of cellophane, and then the top and the bottom sides of the wrapping cellophane were removed precisely (Fig. 2.15).

2.5.6.4 Water Loss (WL) Monitoring

Specimens' weight was monitored during the curing period. After 14 days of curing, specimens are removed from the curing chamber, sealed and tagged in a plastic bag and kept at room temperature until testing (Fig. 2.16).

Fig. 2.16 Specimens sealing until needed for testing

a) b)

Fig. 2.17 ITS (**a**) and ITSM (**b**) devices

2.5.6.5 Testing

Indirect Tensile Strength (ITS)

The indirect tensile strength (ITS) was measured by loading the cylindrical specimens with a compressive load along a vertical diametric plane at a constant rate (Fig. 2.17a).

A constant tensile stress perpendicular to the diametrical plane was developed by the loading configuration and ultimately causing the specimen to fail by the creation of the crack which eventually split the specimen along the vertical diameter. This test was performed at room temperature (25 °C) with a constant displacement rate of 51 mm/minute. The strength of the tested material (S_t) is calculated by the following Eq. (2.12):

$$S_t = \frac{2000 \cdot P}{\pi \cdot t \cdot D} \tag{2.12}$$

where S_t is the tensile strength of the mix (kPa), P describes maximum load, (N), t is the specimen height immediately before the test (mm), and D is the specimen diameter (mm).

Indirect Tensile Stiffness Modulus (ITSM)

To measure the stiffness of the mixes, an indirect tensile stiffness modulus (ITSM) test was performed (Fig. 2.17b). The stiffness modulus is considered as one of the most important performance characteristics of the pavement structures. ITSM tests on cylindrical specimens give information on the viscoelastic behaviour of asphalt mixes. ITSM test measured the stiffness modulus by a series of repeated application of vertical load pulses. The tests are performed in strain control mode with a precise recovery time. According to the standard, the average of 5 load pulses determines the stiffness modulus of the mix. For each pulse, the stiffness modulus is obtained with the following Eq. 2.13:

$$S_m = \frac{F \cdot (R + 0.27)}{L \cdot H} \tag{2.13}$$

where S_m represents the stiffness modulus (MPa), F is the peak value of the applied repeated vertical load (N), H represents mean amplitude of the horizontal deformation obtained from five applications of the load pulse expressed (mm), L is the mean thickness of the specimen (mm), and R is the Poisson's ratio (assumed to be 0.35).

ITSM test was conducted at three different temperatures (2, 10 and 20 C). The modulus also needs to be defined along the perpendicular diameter. It should be 80–110% of the first value to accept the validity of the test. The final value of the modulus is the mean value of both diameters (vertical and horizontal).

2.6 Results and Analysis of ILTP

The main objective of this research work is to better understand the behavior of CBTM, but more precisely, to evaluate the impact of the RA source, the compaction method, the curing methodology and the bitumen source on the properties of the mixes. The results of the ILTP presented here helps to shed lights on many questions related to CBTM. First, the impact of the curing methodology on the water loss is shown. Afterwards, the impacts of the different variables on the mechanical properties studied through different tests are presented.

2.6.1 *Study of the Water Loss (WL)*

In the ILTP, two different curing protocols were used. In one case, free surface dry (FSD), the water can evaporate from all sides of the specimen. For the second protocol, partially sealed drying (PSD), the specimens were wrapped in plastic film and the water can only exit the specimen by the top surface. In both cases, the cure lasted 14 days at 40 °C, and, when possible, the humidity was kept constant at 55%.

The moisture loss was evaluated by weighing the specimens at different times during the 14 days curing at 40 C. A comparison of the water loss for specimens compacted with the SGC (100 mm) and the Marshall hammer is presented in Fig. 2.18. As it can be observed, the majority of the water loss happens in the first three days. It can also be observed that the WL trend does not differ significantly according to the compaction method, but that the Marshall specimens have overall higher water loss. As expected, specimens where water was able to evaporate from all sides, FSD, lose water faster [91].

A third curing protocol was used in one lab to better study the WL. The restricted-surface drying (RSD) in which the specimens are fully sealed was compared to the other two protocols. The results shown in Fig. 2.19 show, as expected, that the RSD results in the slowest water (moisture) loss.

2.7 Summary of the ILTP Results

The result analysis presented in the following sections is based on the properties of the specimens after their 14 days cure. The impact of the curing protocol, so indirectly of the WL, is included in the analysis.

Tables 2.13, 2.14, 2.15 and 2.16 show the average ITS and ITSM results obtained from all the participating laboratories.

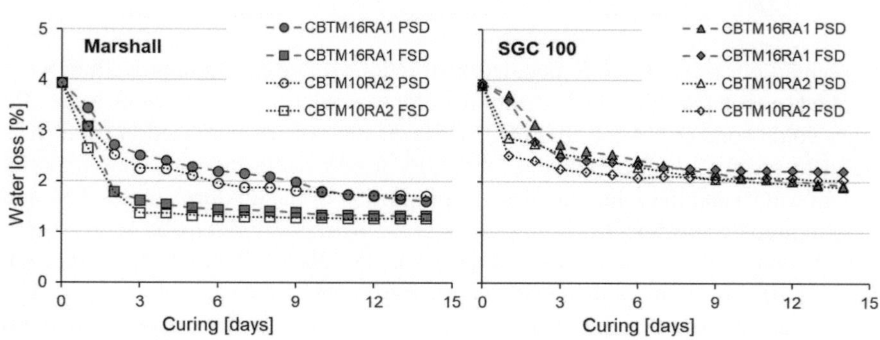

Fig. 2.18 Water loss of two CBTM (0–10 and 0–16 mm) cured for 14 days at 40 °C following the FSD or PSD protocol [91]

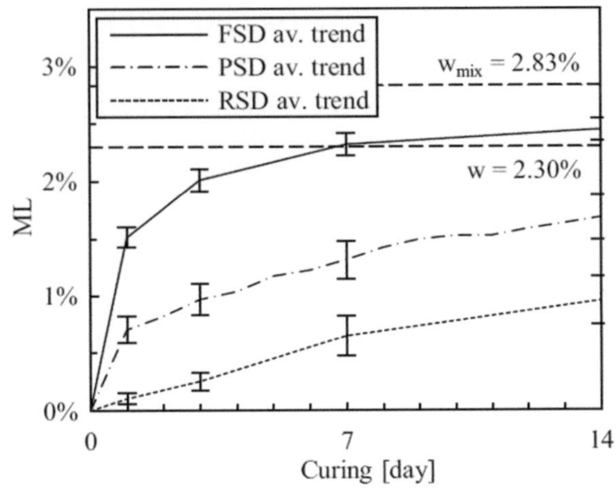

Fig. 2.19 Moisture loss (ML) of CBTM cured for 14 days at 40 °C following the FSD, PSD or RSD protocol [92]

The testing program collected a large number of samples. In order to make an informed judgment, we wanted to compare population means rather than sample means through confidence intervals as much as possible. In doing so, it is assumed as a first approximation that the results have a normal distribution. In the course of the analysis, the plotting of histograms allows the validation of this hypothesis. Nevertheless, in some cases, the distribution of the results is not normal. However, we have treated these results according to a normal distribution, keeping in mind the potential precariousness of this choice in the analysis. Moreover, in some of these cases, we have limited the analysis of the results in the perspective of the samples by referring essentially to the mean and the standard deviation of the samples.

2.7.1 Voids Content Variability in Respect to the Compaction Type

As pointed out in Sect. 2.6.5, detailed instructions on compaction procedure were supplied to each participant laboratory in order to produce specimens with a specific target voids content (14.3% ± 1.0%) whatever the type of compaction used. It is useful to recall that all participant laboratory used exactly the same raw material (except water) and the same mix design (proportion of the mixture: Tables 2.11 and 2.12) to prepare specimens.

In this inter-laboratory test program, as many as 230 specimens were compacted. Figure 2.20 shows the voids content of all specimens compacted in the ILTP. The results clearly show that the objective was (i.e., the production of specimens with substantially the same voids content among the different laboratories using different compaction procedures) was basically achieved.

However, some remarks could be raised analyzing in detail the collected results.

Table 2.13 Summary of the ITS results from each participant laboratory (dry and wet conditions) for CBTM-EMU (average of 3 specimens)

					LAB-A				LAB-B		LAB-C				LAB-D				LAB-E		LAB-F	
					ITS		Voids		ITS		ITS		Voids		ITS		Voids		ITS		ITS	
					Mean kPa	SD kPa	Mean %	SD %	Mean kPa	SD kPa	Mean kPa	SD kPa	Mean %	SD %	Mean kPa	SD kPa	Mean %	SD %	Mean kPa	SD kPa	Mean kPa	SD kPa
CBTM-EMU	DRY	RAP1	FSD	Marshall							443	22	16.7	0.1	957	157	12.2	0.5				
				SGC 100							366	32	14.8	0.9	943	205	12.3	0.5				
				SGC 150	610	76	12.8	0.1	456	60					887	87	12.6	0.1				
				Proctor 150											814	19	12.8	0.4				
			PSD	Marshall							516	30	15.5	0.3	1082	27	10.9	0.6				
				SGC 100							459	7	16	0.5	1046	135	12.1	0.2				
				SGC 150	652	28	13.2	0.40							797	41	13	0.1				
		RAP2	FSD	Marshall											901	76	15	0.4				
				SGC 100											850	50	14.1	0.2				
				SGC 150															711	63		
				Proctor 150																		
			PSD	Marshall																		

(continued)

Table 2.13 (continued)

					LAB-A				LAB-B		LAB-C				LAB-D				LAB-E		LAB-F	
					ITS		Voids		ITS		ITS		Voids		ITS		Voids		ITS		ITS	
					Mean kPa	SD kPa	Mean %	SD %	Mean kPa	SD kPa	Mean kPa	SD kPa	Mean %	SD %	Mean kPa	SD kPa	Mean %	SD %	Mean kPa	SD kPa	Mean kPa	SD kPa
				SGC 100																	611	73
	WET	RAP1	FSD	Marshall											852	88	12	2.3				
				SGC 100											824	96	12.7	0.5				
				SGC 150	607	37	13	0.2	435	70					710	46	12.2	0.3				
				Proctor 150											606	70	13	0.4				
			PSD	Marshall											833	98	11.4	0.7				
				SGC 100											1007	58	12.9	0.6				
				SGC 150	627	33	13.9	1.5							622	34	13.2	0.1				
		RAP2	FSD	Marshall											858	99	14.6	0.1				
				SGC 100											759	112	14.4	0.3				
				SGC 150															574	48		
			PSD	SGC 100																	535	8

Table 2.14 Summary of the ITS results from each participant laboratory (dry and wet conditions) for CBTM-FB (average of 3 specimens)

					LAB-A		LAB-C				LAB-D				LAB-E		LAB-F	
					ITS		ITS		Voids		ITS		Voids		ITS		ITS	
					Mean kPa	SD kPa	SD kPa	SD	Mean %	SD %	Mean kPa	SD kPa	Mean %	SD %	Mean kPa	SD kPa	Mean kPa	SD kPa
CBTM-FB	DRY	RAP1	FSD	Marshall			720	114	11.7	0.8	540	54	17.8	1.2				
				SGC 100			390	53	15.1	0.5	683	66	16.4	0.2				
				SGC 150	688	96					562	39	16.4	0.6	483	3		
				Proctor 150			173	26	21.4	2	520	115	0	0				
			PSD	Marshall			905	50	11.2	0.5	786	100	18	0.6				
				SGC 100			384	50	15.1	0.4	779	13	15.7	0				
				SGC 150							684	62	15.2	0.2				
				Proctor 150			191	40	20.3	2.6	676	43	0	0				
		RAP2	FSD	Marshall							512	37	17.7	0.7				
				SGC 100							670	30	13.8	0.3				
				SGC 150											483	44		
			PSD	Marshall							616	78	17.3	0.4				

(continued)

Table 2.14 (continued)

					LAB-A		LAB-C				LAB-D				LAB-E		LAB-F	
					ITS		ITS		Voids		ITS		Voids		ITS		ITS	
					Mean kPa	SD kPa	SD kPa	SD	Mean %	SD %	Mean kPa	SD kPa	Mean %	SD %	Mean kPa	SD kPa	Mean kPa	SD kPa
				SGC 100							1037	24	11.7	0.1			627	17
	WET	RAP1	FSD	Marshall							360	89	17.1	0.2				
				SGC 100							539	139	15.9	0.1				
				SGC 150	463	42					407	14	16.4	0.4				
				Proctor 150							418	136	0	0				
			PSD	Marshall							573	13	17.3	0.3				
				SGC 100							672	57	15.9	0.4				
				SGC 150							685	63	14.9	0.3				
				Proctor 150							634	158	0	0				
		RAP2	FSD	Marshall							432	22	16.6	0.6				
				SGC 100							567	86	13.2	0.1				
				SGC 150											459	9		
			PSD	Marshall							586	24	17.4	0.5				
				SGC 100							827	71	13.2	0.2			576	70

Table 2.15 Summary of the ITSM results from each participant laboratory for CBTM-EMU (average of 3 specimens)

					LAB-A				LAB-D				LAB-F	
					ITSM		Voids		ITSM		Voids		ITSM	
				T-essai (°C)	Mean kPa	SD kPa	Mean %	SD %	Mean kPa	SD kPa	Mean %	SD %	Mean kPa	SD kPa
CBTM-EMU	RAP1	FSD	Marshall	2					9249	827	12.6	0.5		
				10					7224	739	12.6	0.5		
				20					5972	614	12.6	0.5		
			SGC 100	2					8570	666	12.7	0.5		
				10					7006	605	12.7	0.5		
				20					4907	1053	12.7	0.5		
			SGC 150	2					8295	2387	13	0.1		
				10					7736	1944	13	0.1		
				20	5188[a]	985	12.9	0.30	7603	1833	13	0.1		
			Proctor 150	2					10634	516	13.3	0.4		
				10					8649	419	13.3	0.4		
				20										
		PSD	Marshall	2					9549	418	11.3	0.6		
				10					8290	581	11.3	0.6		
				20					6349	544	11.3	0.6		
			SGC 100	2					9872	795	12.7	0.5		
				10					8279	753	12.7	0.5		
				20					6554	332	12.7	0.5		
			SGC 150	2					6234	1949	13.4	0.1		

(continued)

Table 2.15 (continued)

					LAB-A				LAB-D				LAB-F	
					ITSM		Voids		ITSM		Voids		ITSM	
				T-essai (°C)	Mean kPa	SD kPa	Mean %	SD %	Mean kPa	SD kPa	Mean %	SD %	Mean kPa	SD kPa
				10					6805	1599	13.4	0.1		
				20	5074[a]	375	13.8	1.30	7176	1969	13.4	0.1		
	RAP2	FSD	Marshall	2					7001	1563	16.9	0.4		
				10					6709	1847	16.9	0.4		
				20					6454	1718	16.9	0.4		
			SGC 100	2					8431	415	16.2	0		
				10					6193	156	16.2	0		
				20					4344	271	16.2	0		
		PSD	SGC 100	2									4739[b]	247
				10									3991	236
				20									3012	422

[a] Test done à 25 °C
[b] Test done à 0 °C

Table 2.16 Summary of the ITSM results from each participant laboratory for CBTM-FB (average of 3 specimens)

					LAB-A				LAB-D				LAB-F	
					ITSM		Voids		ITSM		Voids		ITSM	
				T-essai (°C)	Mean kPa	SD kPa	Mean %	SD %	Mean kPa	SD kPa	Mean %	SD %	Mean kPa	SD kPa
CBTM-EMU	RAP1	FSD	Marshall	2					9249	827	12.6	0.5		
				10					7224	739	12.6	0.5		
				20					5972	614	12.6	0.5		
			SGC 100	2					8570	666	12.7	0.5		
				10					7006	605	12.7	0.5		
				20					4907	1053	12.7	0.5		
			SGC 150	2					8295	2387	13	0.1		
				10					7736	1944	13	0.1		
				20	5188[a]	985	12.9	0.30	7603	1833	13	0.1		
			Proctor 150	2					10634	516	13.3	0.4		
				10					8649	419	13.3	0.4		
				20										
		PSD	Marshall	2					9549	418	11.3	0.6		
				10					8290	581	11.3	0.6		
				20					6349	544	11.3	0.6		
			SGC 100	2					9872	795	12.7	0.5		
				10					8279	753	12.7	0.5		
				20					6554	332	12.7	0.5		
			SGC 150	2					6234	1949	13.4	0.1		

(continued)

Table 2.16 (continued)

					LAB-A				LAB-D				LAB-F	
					ITSM		Voids		ITSM		Voids		ITSM	
				T-essai (°C)	Mean kPa	SD kPa	Mean %	SD %	Mean kPa	SD kPa	Mean %	SD %	Mean kPa	SD kPa
				10					6805	1599	13.4	0.1		
				20	5074[a]	375	13.8	1.30	7176	1969	13.4	0.1		
	RAP2	FSD	Marshall	2					7001	1563	16.9	0.4		
				10					6709	1847	16.9	0.4		
				20					6454	1718	16.9	0.4		
			SGC 100	2					8431	415	16.2	0		
				10					6193	156	16.2	0		
				20					4344	271	16.2	0		
		PSD	SGC 100	2									4739[b]	247
				10									3991	236
				20									3012	422

[a] Test done à 25 °C
[b] Test done à 0 °C

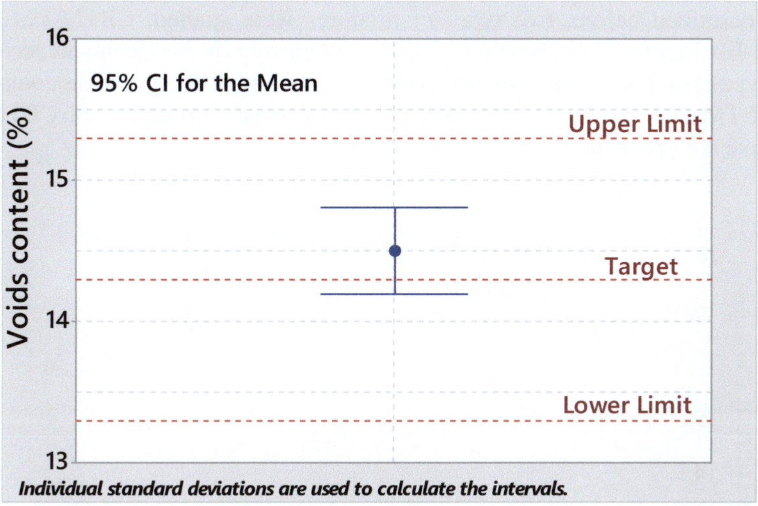

a)

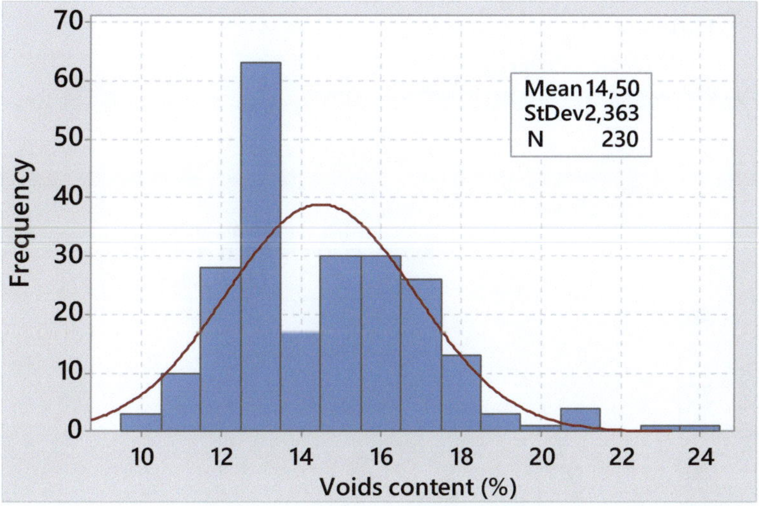

b)

Fig. 2.20 Overall voids content of specimens produced in the ILTP (N = 210; $\bar{x}$ = 14.3%): **a** interval plot, **b** histogram of values. Data from ITS-Dry, ITS-Wet and ITSM specimens

As described before, two types of mixtures were studied: CBTM-EMU and CBTM-FB. Figure 2.21 shows the distribution of air voids in the specimens according to the type of mix. The results show that the voids content of the specimens compacted with CBTM-EMU is below the lower limit set while those compacted with CBTM-FB are above the upper limit set.

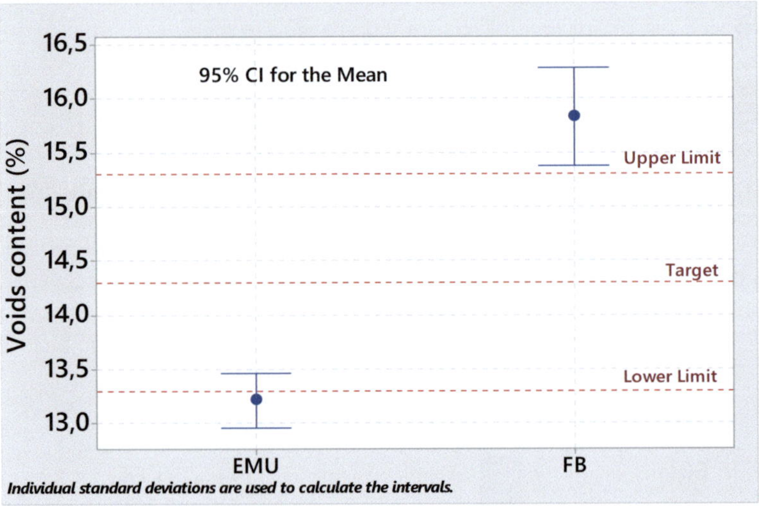

a)

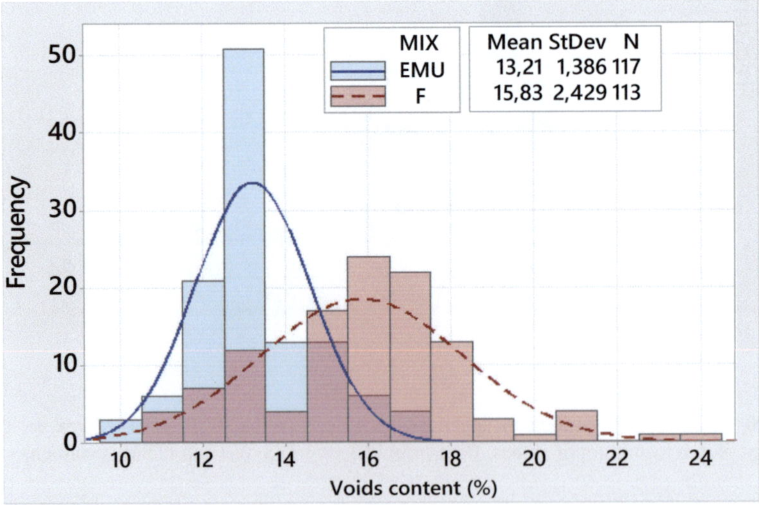

b)

Fig. 2.21 Specimen voids content produced in the ILTP in accordance with the type of mix: **a** interval plot, **b** histogram values. Data from ITS-Dry, ITS-Wet and ITSM specimens

The higher air voids observed for the FB mixes could be due to the stability of the foam. For most labs, the mixes were produced in different batches that were mixed together before compaction, which takes time. One hypothesis could be that the slow setting emulsion remains stable longer than the foam which could facilitate the compaction. However, the voids content variation could also be attributed to laboratories operations.

To verify if the deviations are attributable to the laboratory operations, Fig. 2.22a) presents the variation of voids content in the compacted specimens according to the type of mixtures and the laboratory concerned. The results in Fig. 2.22a indicates that independently of the laboratory, the voids content of the specimens is largely different according to the type of mixtures. Figure 2.22b also shows that regardless of the type of compaction recommended, compaction is always more difficult with the CBTM-FB mix.

Figure 2.23 shows the overall voids content of compacted specimens in accordance with the compaction type used. The histograms presented in Fig. 2.23a show a normal distribution of results. It should be noted that results from Proctor compactor (Pro) show a tighter distribution, but this could be explained by a very limited number of specimens (N = 6).

From Fig. 2.23b, it should be noted that the SGC 100 mm results (SGC10) are the closest to the target and that the voids content from the SGC 100 mm (SGC10), SGC 150 mm (SGC15) and Marshall (M) are not statistically different. Considering all these results presented in Fig. 2.23b, the confidence interval of the means of a normal distribution for each of the compaction methods studied in this ILTP exchange program were calculated and are given in Table 2.17 and in Fig. 2.22b. The confidence interval is set at 95%.

2.7.1.1 Voids Content Variability in Respect to the CBTM-EMU

Figure 2.24 report the interval plot of the voids content as regards to the type of compaction when focusing on the mix type CBTM-EMU as reported by laboratories A and D.

Figure 2.24 shows that the means for CBTM-EMU mixes are clearly below to the lower limit target for the testing program. This is somewhat surprising for SGC specimens for which voids content is normally easy to control. One possible explanation was introduced in Sect. 2.4.4 where cement hydrate products accounted in the global balance of volumetric components shows an apparent reduction of the volume of voids of compacted CRM (V_m) in case of complete cement hydration process. Based on that and referring to Eq. 2.9a (Sect. 2.4.4) and with assuming a complete hydration of the cement ($\alpha = 1$), for a target V_m^0 equal to 14.3% as properly targeted especially for SGC specimens, the $V_m^{\alpha=1}$ must be equal to 13.3%, value as calculated in accordance with Eq. 2.9a. This value is closer to data reported on Fig. 2.24.

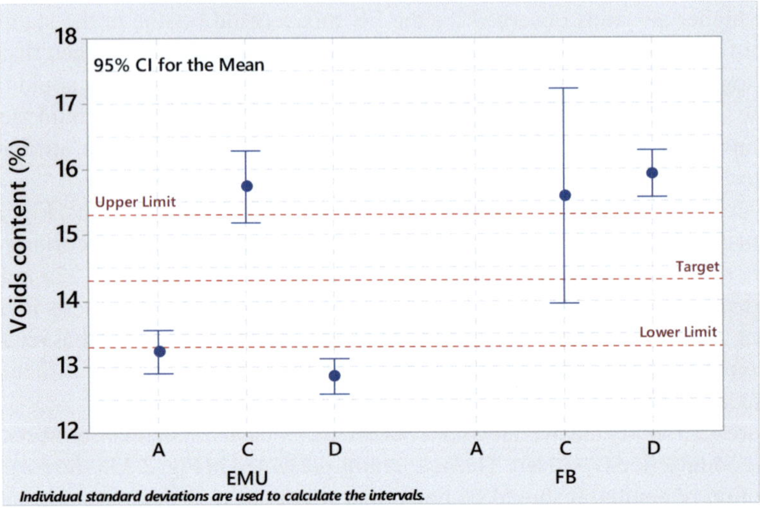

a)

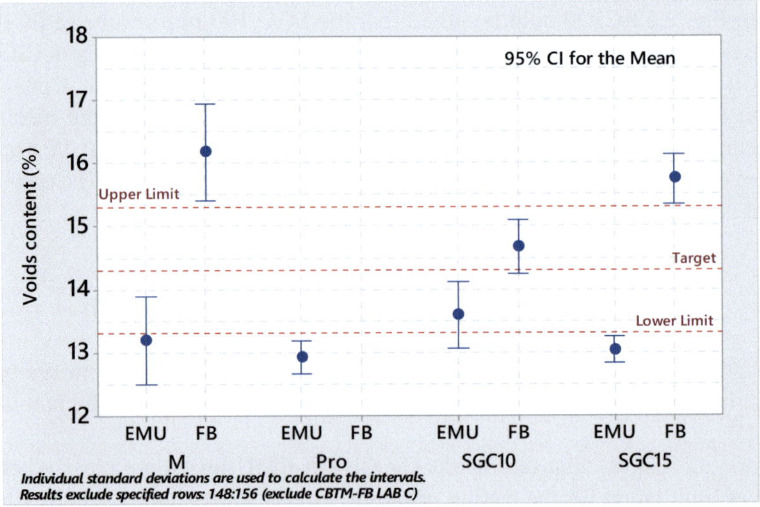

b)

Fig. 2.22 Interval plots of specimens' voids content produced in the ILTP in accordance with the type of mix, the laboratory and the compaction type: **a** laboratory and type of mix, **b** type of mix and compaction type. Data from ITS-Dry, ITS-Wet and ITSM specimens

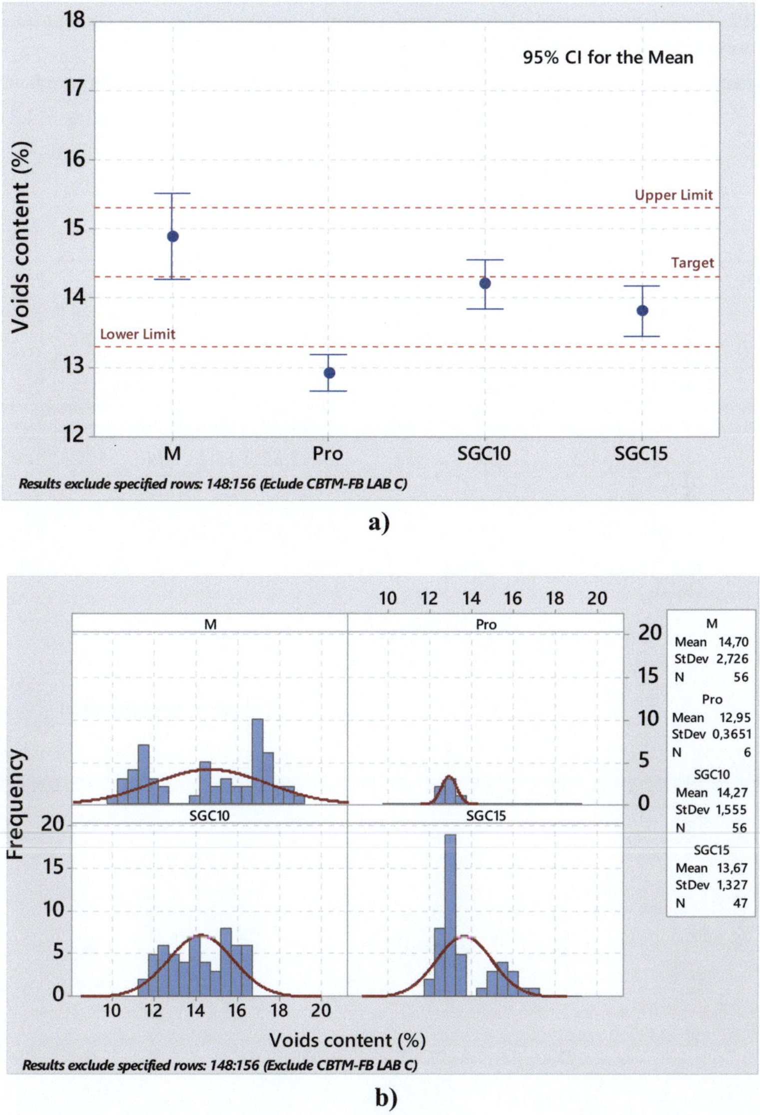

Fig. 2.23 Voids content in accordance with compaction types: **a** interval plot, **b** histogram values. Data from ITS-Dry, ITS-Wet and ITSM specimens

2.7.1.2 Voids Content Variability in Respect to the CBTM-FB

As regards CBTM-FB results, it appears that the compaction procedure provided to each participant laboratory has not permitted to control the air voids level with enough

Table 2.17 Confidence interval of the mean of a normal distribution of voids content for each of the compaction methods studied (95% confidence interval for the mean)

Compaction type	Mix type	Specimen mean	Standard deviation	Specimen size	Interval of the population mean*		Meet requirements?	
					Upper limit	Lower limit	14.3 ± 1%	14.3 ± 2%
		$\overline{x}_i$	$StDev_i$	N_i	$\overline{X}_i^+$	$\overline{X}_i^-$		
Marshall	FB	16.16	2.50	44	15.40	16.92	No	Partly
	EMU	13.18	1.98	33	12.48	13.88	No	Yes
SGC 100	FB	14.66	1.43	44	14.22	15.09	Yes	Yes
	EMU	13.58	1.50	33	13.04	14.11	Partly	Yes
SGC 150	FB	15.73	0.76	17	15.34	16.12	No	Yes
	EMU	13.02	0.66	42	12.82	13.23	No	Yes
Proctor	FB	–	–	–	–	–	–	–
	EMU	12.91	0.35	9	12.65	13.18	No	Yes

* $\overline{X}_i = \overline{x}_i \pm \frac{t_{(\alpha/2)}}{\sqrt[2]{N_i - 1}} \cdot StDev_i = \left[\overline{X}_i^- ; \overline{X}_i^+\right]$

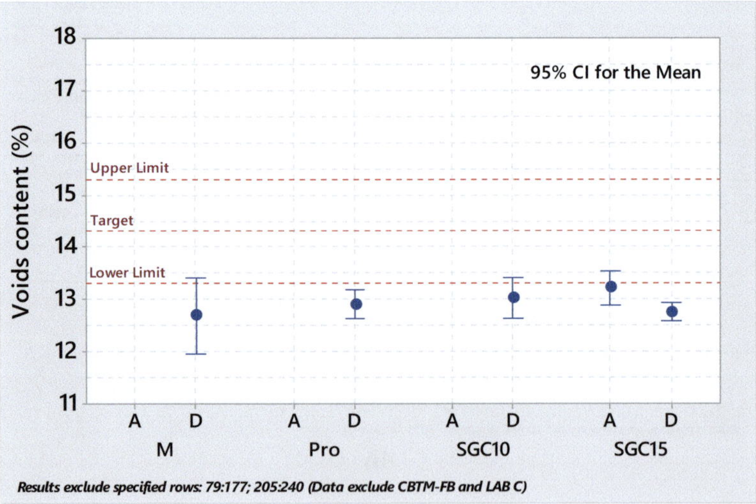

Fig. 2.24 Interval plot of voids content for LAB-A and LAB-D and type of compaction (Only CBTM-EMU data)

precision. One reason behind that is, as mentioned before, the way to produce CBTM-FB in laboratories. Raschia et al. [45] addressed the inter-batch variability related to the laboratory production technique of CBTM-FB. Indeed, because the batch size required to produce all the specimens to be compacted with all the selected compaction methods was too high, many batches (8 for LAB-D; B1 to B8) were mixed in the laboratory to provide all the specimens. Figure 2.25 shows the variation in maximum specific gravity and the variation in bitumen content for each of the batches produced by LAB-D.

Although the mass of the constituents was carefully controlled, it remains that the procedure used to dose the foamed bitumen content per batch brings a fluctuation in the volumetric proportions of the constituents (Fig. 2.25). Raschia et al. [45] brings interesting elements of reflection on the subject.

Nevertheless, the results on Fig. 2.25 show that the B4 batch presents both a bitumen content and a maximum density (G_{mm}) comparable to the theoretical values established according to the proportions of the mixture fixed and the properties of the materials. For batch B4, 9 specimens were compacted with the SGC (SGC-150) and the average value of the voids is 15.1% with a standard deviation of 0.2% as report in Fig. 2.26.

2.7.1.3 Voids Content Variability for SGC Compacted Specimens

The gyratory shear compactor (SGC) is commonly used to evaluate the workability in the laboratory of conventional asphalt mixtures (HMA). The Marshall and Proctor compaction methods are also commonly used internationally to evaluate the workability of CBTM. However, CBTMs are laid on site with the same compaction tools as HMAs. Although the presence of water in CBTM-EMU can interfere with the pore pressure compaction process, Grilli et al. [15] have clearly demonstrated the ability of SGC to produce a realistic evaluation of the workability of CBTM in the laboratory. One of the advantages of this compaction method is that it allows to control the volumetric proportions of the compacted CBTM in the laboratory by ensuring a target void content, which significantly impacts the thermo-mechanical properties of CBTM.

Currently, two diameters of specimens are produced at the SGC, 100 and 150 mm and studied in this ILTP. However, the use of relatively coarse RAP (GNM of the order of 14–28 mm) is common to produce CBTM. Figure 2.27 shows the histograms of voids content values of the 100 and 150 mm diameter compaction specimens. According to the results of Fig. 2.27a), it seems that the compaction is more effective for the 150 mm specimens on the one hand, and that the relative variability is slightly higher for the 100 mm specimens i.e., 10.9% compared to 9.7% for the 150 mm specimens. When we discretize the results according to the type of mix (Fig. 2.27b), we see that the average void content and the variability are distinct according to the type of mix, with visibly higher values for CBTM-FB. Considering the values for CBTM-EMU, the void contents of the 150 mm specimens are clearly more centered

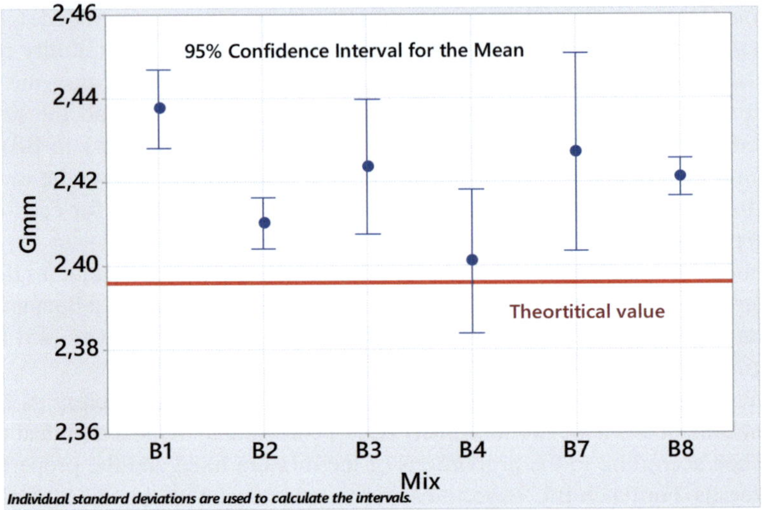

a)

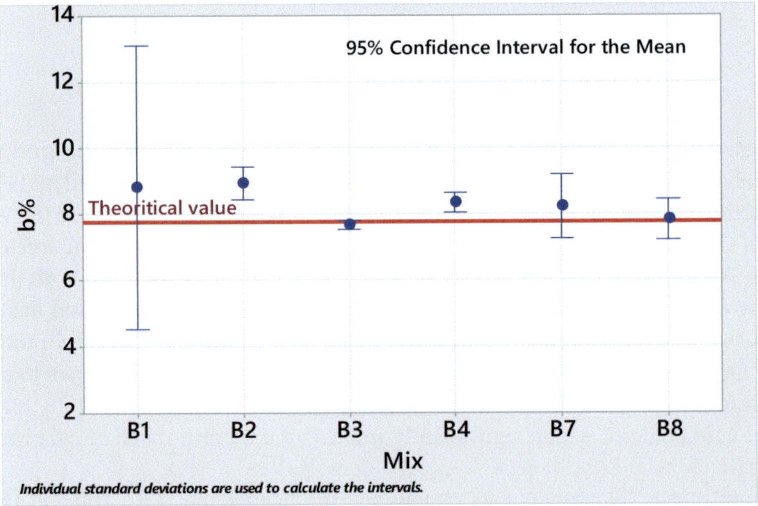

b)

Fig. 2.25 Interval plot of the maximum specific gravity (G_{mm}) and bitumen content (b%) for each CTBM-FB batch mixed by LAB-D: **a** maximum specific gravity, **b** total bitumen content (Data RAP1 only)

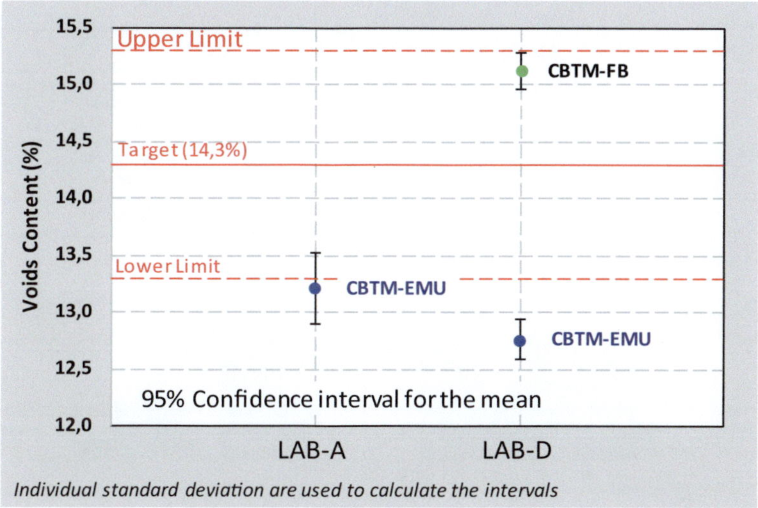

Fig. 2.26 Interval plot of voids content of CBTM-EMU and CBTM-FB from batch B4 compacted with SGC 150

with a variability of about 5%. The production of 150 mm specimens at SGC provides more homogeneous samples.

2.7.1.4 Conclusion on Air Voids Distribution

To have comparable air voids to facilitate comparison between CBTM batches from the different laboratories, a compaction procedure was developed and used by all the labs. The results show that overall, the air voids are well controlled with all the results being with 2% of the targeted value. However, there are difference in the air voids distribution depending on the type of mixes, CBTM-EMU and CBTM-FB, and between the different compaction methods, SGC, Marshall and Proctor. Overall, mixes made with foam have higher voids than mixes with emulsions. As for the compaction, both SGC, 100 and 150, and Marshall give similar air voids and Proctor lower air voids.

2.7.2 RAP Sources Impact on CBTM Mechanical Properties

One of the main objectives of the ILTP is to verify whether the classification of RAP according to the fragmentation test [45] allows a differentiation of CBTM with respect to their mechanical properties. Following the usual approach of CBTM formulation, we refer particularly to the indirect tensile strength (ITS) to evaluate the mechanical performance of CBTM. Their potential sensitivity to water is also

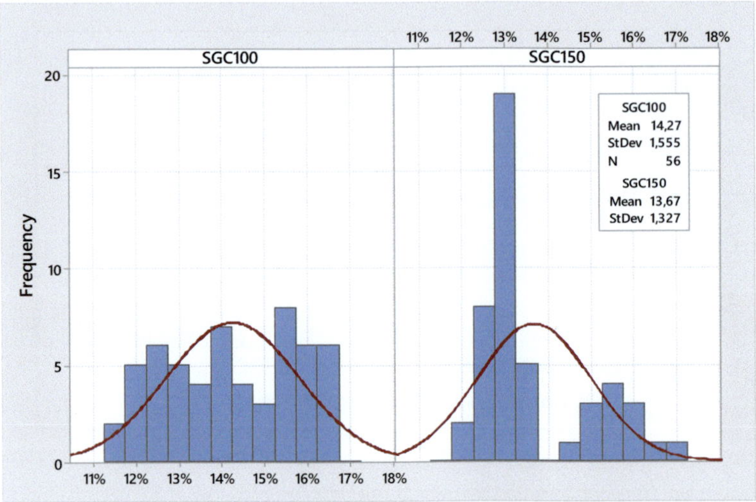

a)

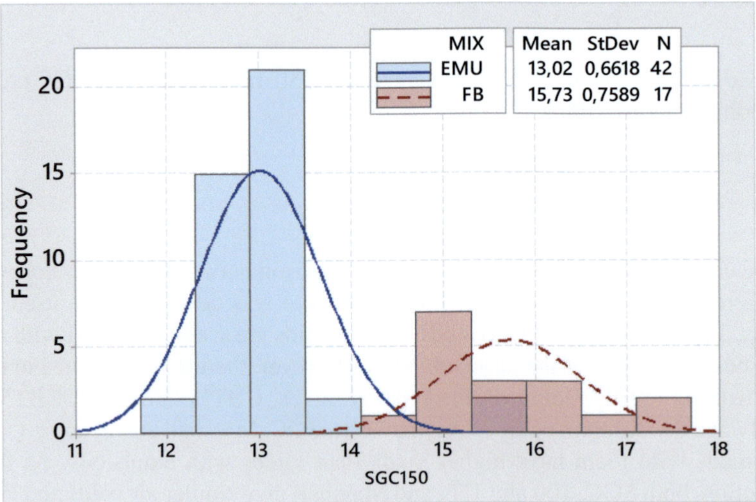

b)

Fig. 2.27 Histogram of SGC voids content in looks to SGC diameter samples: **a** regardless of the mix type, **b** voids distribution for SGC150 mm, **c** voids distribution for SGC100 mm (ITS-Dry, ITS-Wet and ITSMD data)

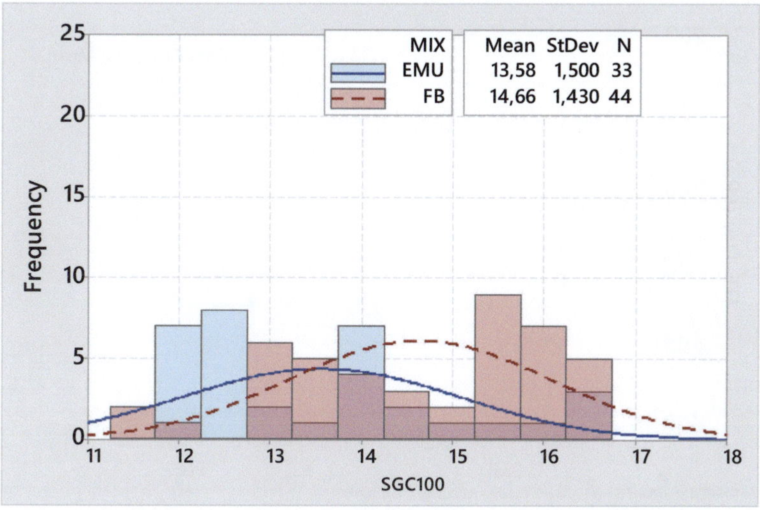

c)

Fig. 2.27 (continued)

evaluated from the ITS, wet vs dry. In addition, several laboratories also refer to the indirect tensile stiffness (ITSM) to judge the thermomechanical properties of CBTM.

2.7.2.1 Indirect Tensile Strength (ITS)

Figure 2.28 shows the confidence intervals of the means of the dry and wet indirect tensile strength (ITS) values of all the tests performed with reference to the RA source. The first thing to note is that the values are high compared to the typical minimum value of 225 kPa used for cold recycling [13]. Overall, the mean values are consistently higher for the specimens incorporating RA2. However, the results in Fig. 2.28 do not show a statistically significant difference in ITS values depending on the source of RA used, for both the dry and wet tested specimens. Obviously, the variation of the voids content in the specimens plays a determining role.

Figure 2.29 shows the histograms of the indirect tensile strengths (ITS) for the tests performed in the dry state (a) and those performed in the wet state (b). The distributions of the results are relatively normal with globally higher average values for the specimens made with RA1. Table 2.18 summarizes the results of Fig. 2.29. The results indicate that the variability is greater for the specimens made with RA1 and tested in the dry state but similar for those tested in the wet state. Moreover, it appears that the water resistance, the indirect tensile ratio (ITR), for the two sources of RA studied is quite acceptable with a noticeably better performance for the RA1.

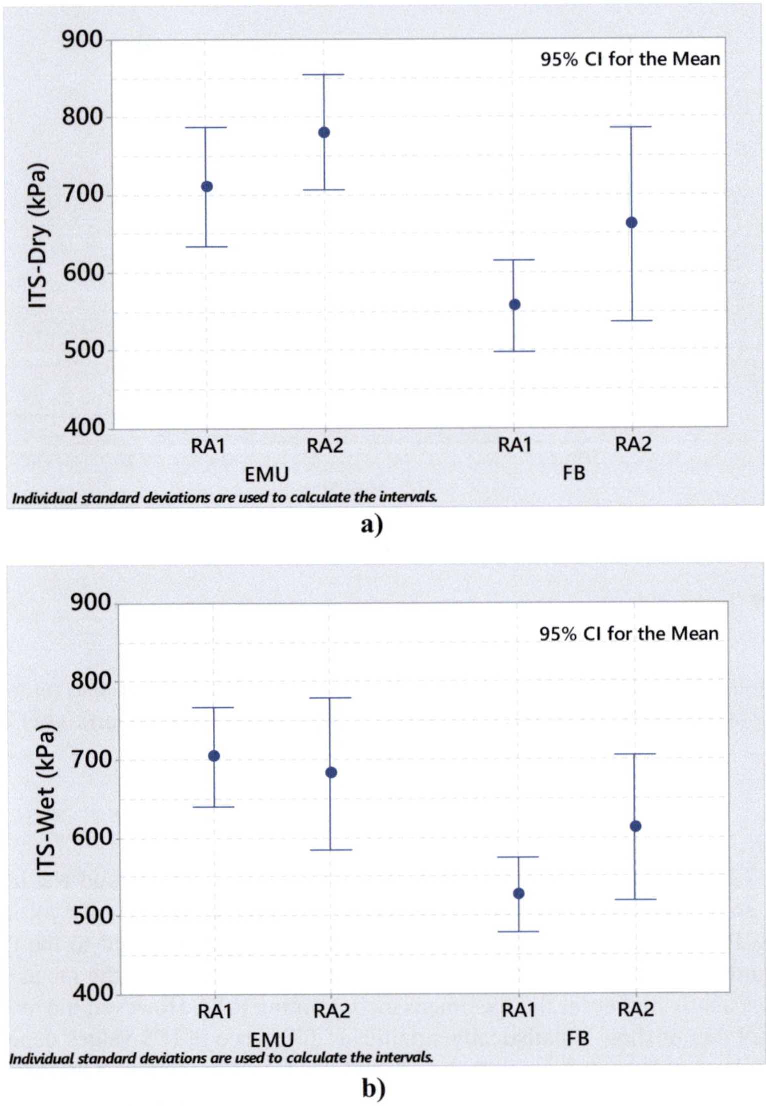

Fig. 2.28 Interval plots of ITS results in accordance with RAP source: **a** ITS-dry (all Data), **b** ITS-wet

Figure 2.30 shows the relationship between the voids content of the specimens and the ITS values. For all the results, an average variation of the ITS value of the order of 58 kPa per percentage of voids can be assumed.

Even if there seems to be a trend between the ITS and the air voids, the variance (R^2) is low so the difference in variation of ITS values by percent voids is not significant between tests performed in the dry or wet condition. Ultimately, although

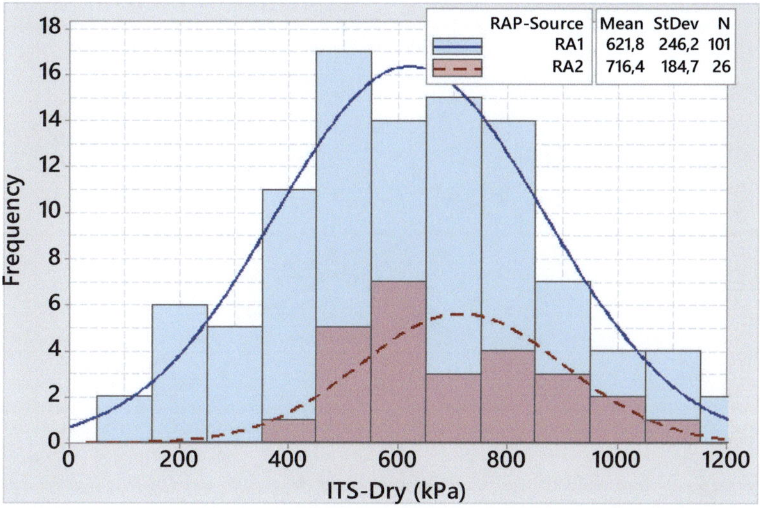

a)

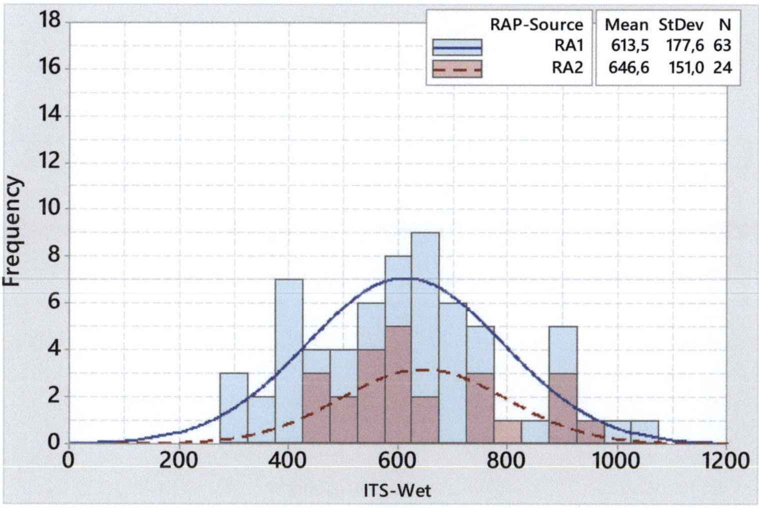

b)

Fig. 2.29 ITS results: RAP source effect: **a** ITS-dry, **b** ITS-wet

Table 2.18 Summary of the results of Fig. 2.29

	Mean values (kPa)		(StDev/Mean) ∗ 100% (%)		ITR (–) (–)
	Dry	Wet	Dry	Wet	
RA1	621	626	40	29	1.01
RA2	707	617	26	23	0.87

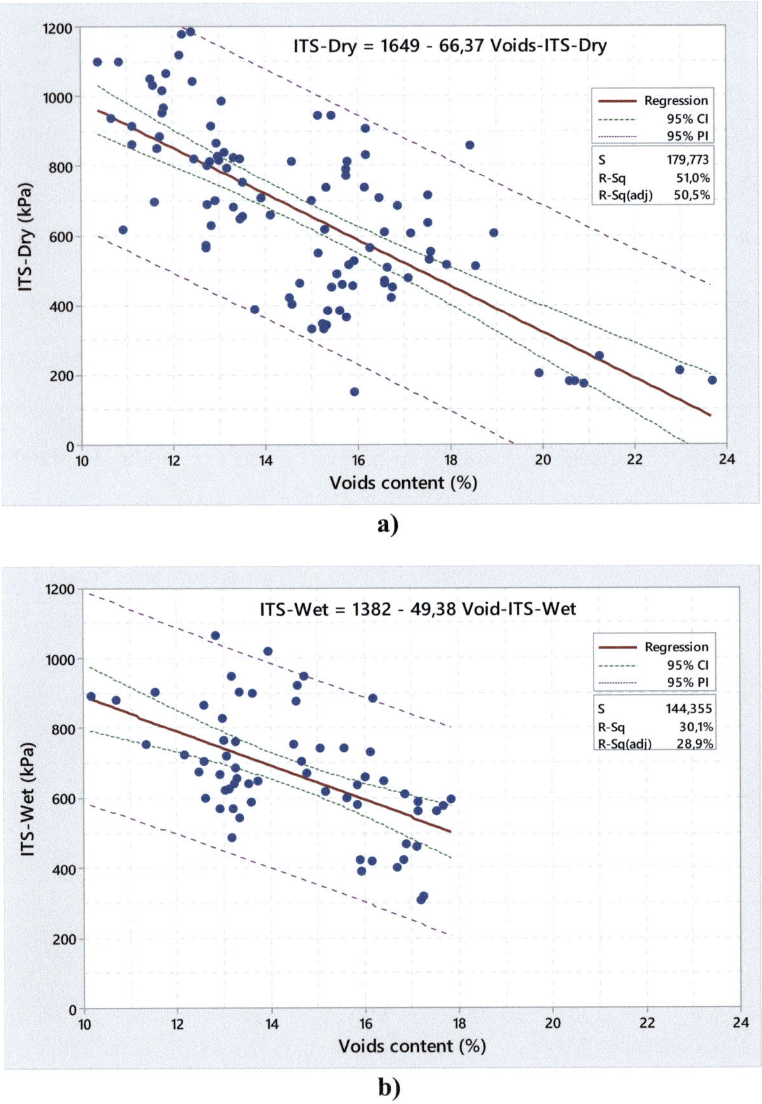

Fig. 2.30 Relationship between ITS results and voids content: **a** ITS-dry, **b** ITS-wet

the previous section highlighted that the voids in the tested specimens are larger for CBTM-FB, the impact of the RA source can be addressed by grouping the results of the two families of CBTM tested as presented in Fig. 2.31a, b. The results show that the ITS-voids relationship is essentially the same regardless of the RA source. Considering all the results and the corresponding confidence intervals fixed at 95% and represented with the help of the transparent surfaces, Fig. 2.31c, d show, on the one hand, that at equivalent voids content, the CBTMs made with RA2 are systematically more resistant than those made with RA1. On the other hand, the two families are statistically distinct. This result is decisive since it suggests the capacity of the fragmentation test to discriminate between the sources of RA in terms of their mechanical property (tensile resistance). However, it is important to underline that the difference in the granularity of the two sources of RA studied and the mix gradation may be the cause of these differences in strength. This aspect is discussed in detail in Sect. 2.6.4.2.

Figure 2.32 shows the deviations according to the type of mixture with respect to the RA source used and the compaction mode used. The results show that, overall, the ITS values are on average higher for CBTM-EMU specimens (~750 kPa) compared to CBTM-FB specimens (~595 kPa). Figure 2.32b, c show the relationship between ITS-Dry values and voids for the Marshall (M) and Gyratory Shear Compactor (SGC10) specimens. Due to the limited number of tests performed for RA2, these figures do not show a statistically significant difference depending on the RA source. Nevertheless, the results show that the average values of the tests with RA2 are always higher than those with RA1.

2.7.2.2 Influence of the Granularity on the Indirect Tensile Strength

The type of RA (RA1 and RA2) does not seem to influence the strength, significantly. The results showed that even if RA2 has fragmentation value lower than RA1, the correction of the gradation curve through mineral filler following the respective maximum density gradation (Fuller distribution) allows CBTM mixture with RA2 to behave similarly to the one with RAP1 [91]. As a parallel work of this TG1, Raschia et al. [40] showed the effect of the two RA used by modifying the grading distribution in order to remove this variable on the mechanical properties. As a result, two mixtures were obtained and produced following the same mix design and protocol, in which the only variable was the RA aggregate. Results showed that RA source did not influence stiffness modulus (ITSM at 3 temperatures), but it was observed that RA1 (with higher fragmentation resistance) conferred to the CBTM material higher ITS resistance, compared to RA2. A following study based on the same concepts and that aimed to study the influence of RA source by removing the different gradation variable was published by Raschia et al. [93]. In this study it was observed that RA1 and RA2, with same gradation, had the same compaction ability by investigating the particles shape and applying the Compressive Packing Model. However, CBTM mixtures produced with RA1 and RA2 were characterized by significantly different rheological properties, especially linked to the aggregate skeleton.

2.7.2.3 ITSM

Figure 2.33 shows the average stiffness values obtained for all the tests at different temperatures considering RA source. At first sight, the results show that the difference

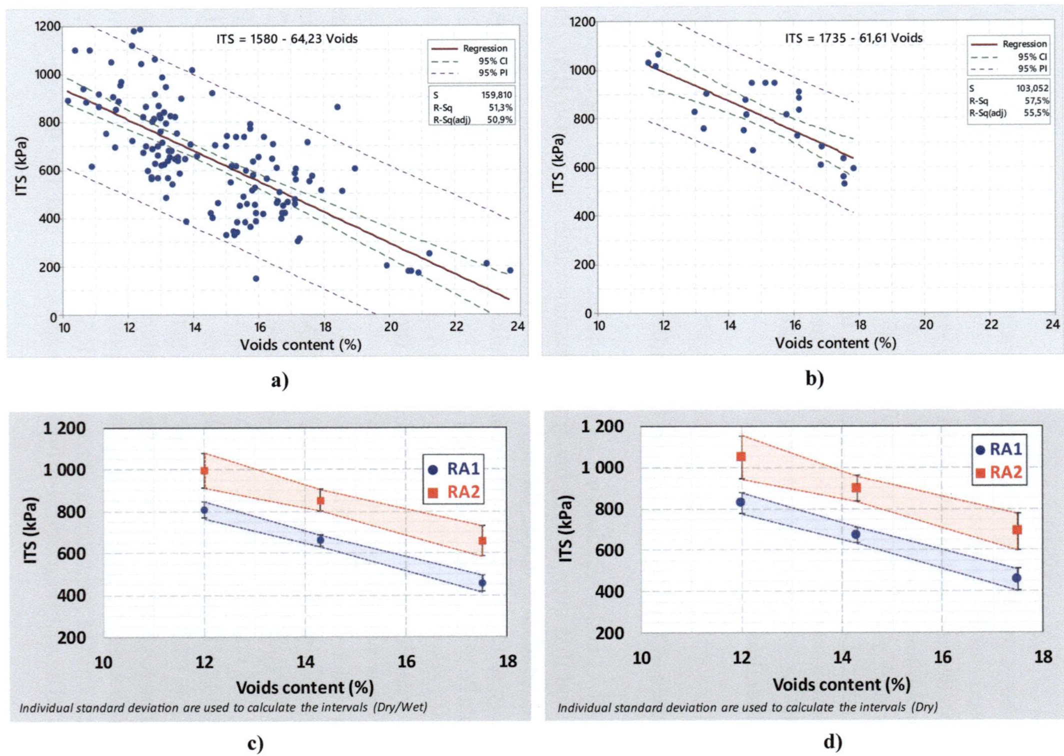

Fig. 2.31 Relationship between ITS results and voids content: **a** RA1: dry and wet data combined, **b** RA2: Dry and wet data combined, **c** RA1 & RA2: dry and wet data, **d** RA1 & RA2: dry data only

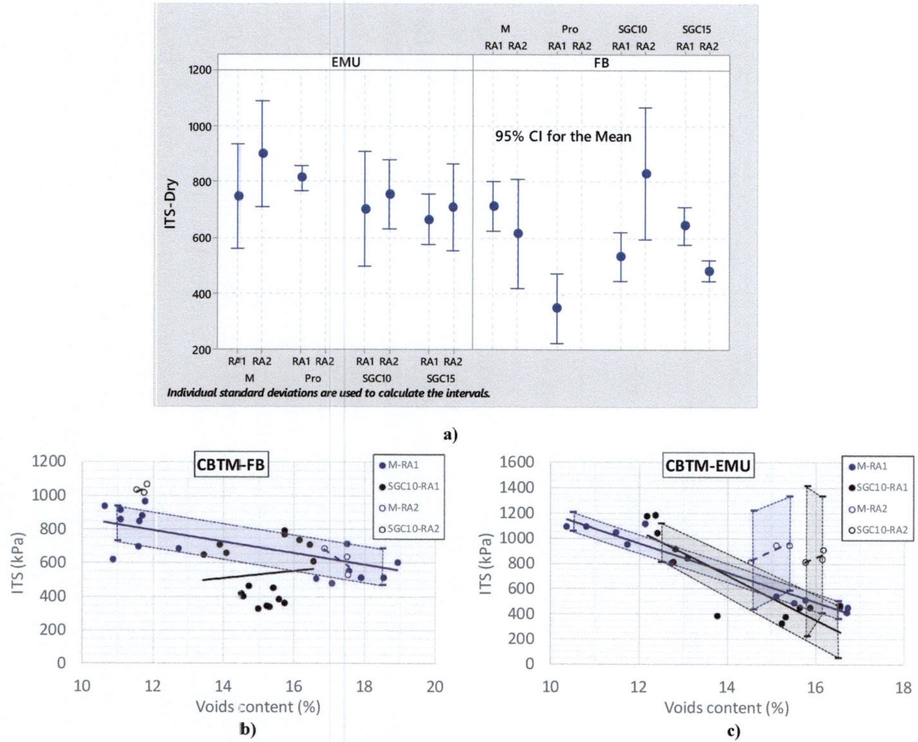

Fig. 2.32 RA source effect in accordance the mix and the compaction types: **a** interval plots of ITS-dry (data of all labs), **b** relationship between ITS-dry and voids for CBTM-FB (M and SGC10 only), **c** relationship between ITS-dry and voids for CBTM-EMU (M and SGC10 only)

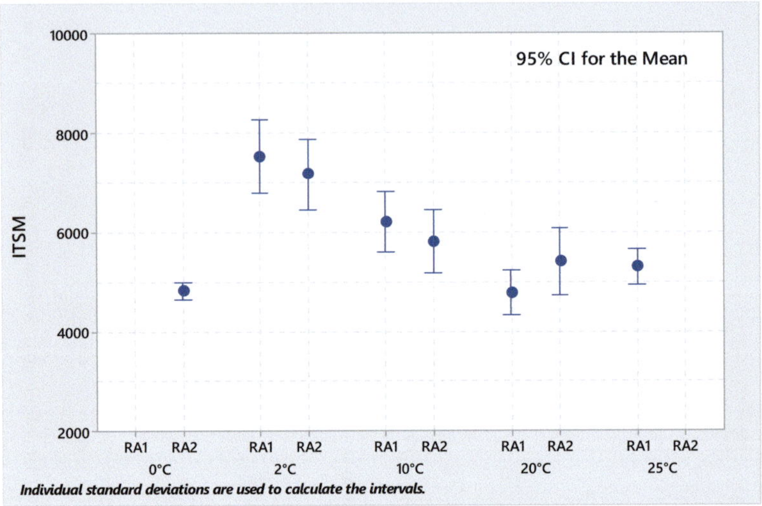

Fig. 2.33 Interval plot of ITSM results in respect to the RA source and the testing temperature

in stiffness according to the RA source is not significant independently of the test temperature.

However, as mentioned in Sect. 2.6.3, the voids content can vary from one specimen to another. Figure 2.34 shows the relationship between the stiffness and voids content of the specimens according to the different mixtures tested (CBTM-EMU vs. CBTM-FB).

An ANOVA analysis confirms that the variation in the relationship between stiffness and voids is independent of the RA source for the CBTM-EMU mixture, but not for the CBTM-FB mixture. Therefore, for the CBTM-EMU mixture, the averages of all the results can be directly compared to evaluate the impact of the RA source. Similarly, since the same specimen was tested at three temperatures (2, 10 and 20 °C), the results can also be analyzed independently of temperature. For CBTM-FB data analysis we must keep in mind that the voids content impact is more effective for RA2.

Figure 2.35 shows the differences in voids content according to the type of mix (a) and according to the compaction mode (b). Globally, the voids content is more important for CBTM-FB specimens (average deviation of more than 3%, i.e., a coefficient of variation-CV of about 20–24%). Knowing that the voids content has a direct impact on the stiffness value (see slope Fig. 2.34), it is appropriate to compare the stiffness values according to the RA source by referring to specimens with more grouped voids content. Figure 2.35b shows the confidence intervals according to the type of compaction used. The deviations according to the type of mix are less when referring to the specimens compacted with the SGC going from 3.1 to 2.1% for an average coefficient of variation lower than 15%. The average values of voids content

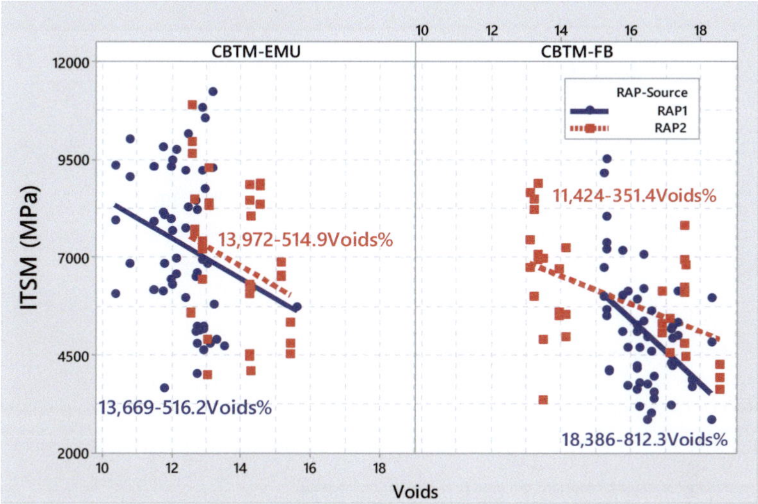

Fig. 2.34 Scatterplot of ITSM-mean values from all results in accordance with the specimens' voids content

for these groups of specimens are respectively 13.0 and 15.1% for CBTM-EMU (CV of 6.2% with N = 27) and CBTM-FB (CV of 8.6% with N = 18).

Figure 2.36 shows in more detail the variation of voids within the SGC specimens depending on the type of mix and the source of RA. The variation in void content according to the type of RA is significant. As reported in Fig. 2.36, the difference in voids content between RA1 and RA2 mixes for SGC10 samples is reversed depending on the type of mix, i.e., +1.9% and −2.6% for CBTM-EMU and CBTM-FB respectively.

To compare on a common basis the stiffness values according to the source of RA, it is necessary to correct the average values and those of the intervals according to the voids content of the specimens. The analysis here is limited to the results of 100 mm specimens, compacted with SGC (SGC10).

Figure 2.37 shows the evolution of the stiffness of 100 mm specimens (SGC10) according to the source of RA and the type of mixture. At first sight, all the ITSM values of CBTM-EMU RA2 are lower than those of CBTM-EMU RA1 at corresponding temperatures. On the other hand, statistically significant differences depending on the RA source are observable for CBTM-EMU at 10 and 20 °C. However, to establish a fair comparison, it is necessary to consider the difference in voids content and to correct the average stiffness obtained, accordingly.

Figure 2.38 shows the variation of the overall ITSM values according to the voids content and the type of mixture. It can be seen that an increase in the voids content of a specimen systematically reduces its stiffness. The increasing regression lines for the CBTM-EMU RA1 results, referring to the SGC10 specimens, are not realistic according to the authors. They are explained by the fact that the variation in voids content of the specimen sample is small, of the order of one percent, which makes

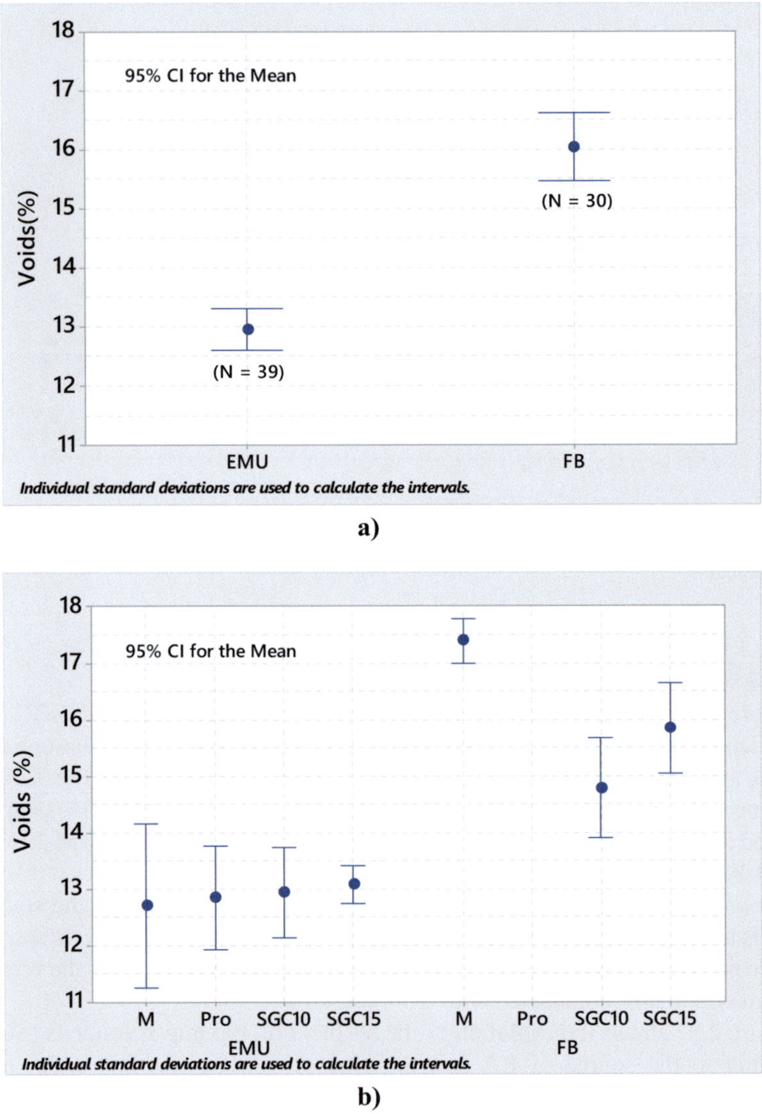

Fig. 2.35 Interval plot of voids content of ITSM specimens: **a** mix type effect, **b** mix and compaction types effect

it impossible to establish a representative stiffness-voids content relationship: these regression lines should be essentially horizontal.

Considering voids content differences (Fig. 2.36) and the correction attributable to it, as compiled in Fig. 2.38, the corrected average ITSM values for CBTM as regards SGC10 data are summarized in Table 2.19 and reported in Fig. 2.39.

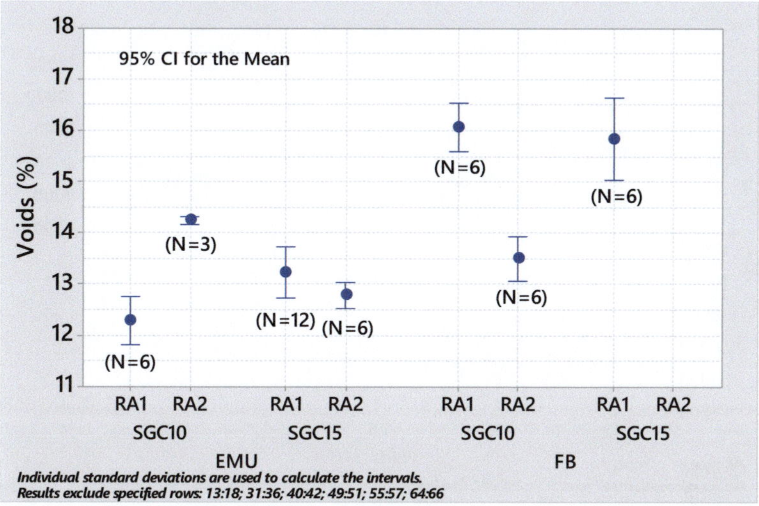

Fig. 2.36 Interval plot of voids content of ITSM specimens obtained from SGC with respect to the RA source

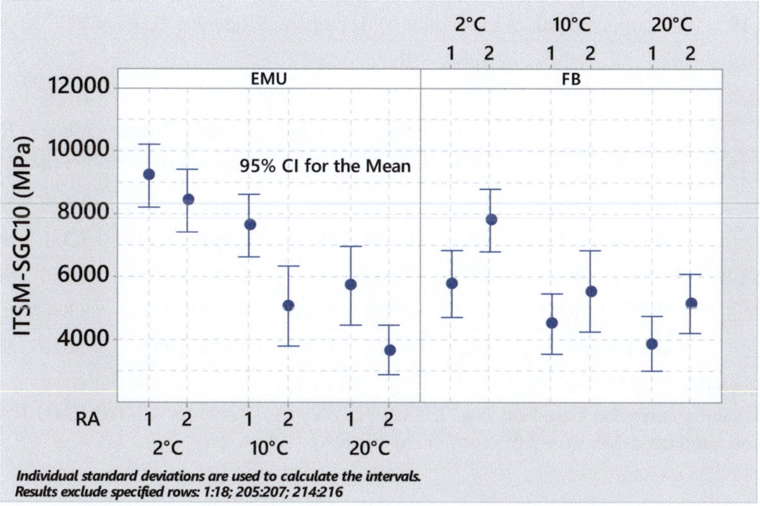

Fig. 2.37 Interval plot of ITSM results on specimens obtained from SGC with respect to the RA source

In Fig. 2.39 the mean and interval corrected values have been slightly shifted along the temperature axis only to allow the impact of the corrections to be clearly visualized. What appeared in Fig. 2.37 to be a significant difference in the stiffness of the CBTMs with respect to the source of RA, the corrected values (Fig. 2.39) following the gaps show instead a clear tightening of the ITSM values according to

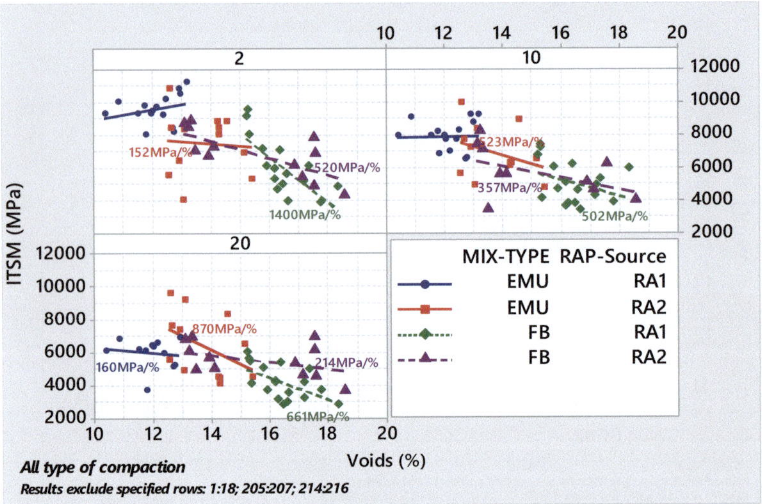

Fig. 2.38 Relationship between voids content and ITSM values for all tested specimens as regards RA source and mix type

Table 2.19 ITSM corrected values to account voids content differences (Data of SGC10 only)

			Voids (%)	Temperature					
				2 °C		10 °C		20 °C	
				Mean	Range	Mean	Range	Mean	Range
CBTM-EMU	ITSM data[a]	RA1	12.3	9,220	1,004	7,642	999	5,730	1,244
		RA2	14.2	8,431	997	5,092	1,269	3,678	804
	RA2-Corr[b]		***12.3***	**8,720**	**997**	**6,086**	**1,269**	**5,331**	**804**
CBTM-FB	ITSM data[a]	RA2	13.5	7,806	998	5,542	1,286	5,158	943
		RA1	16.1	5,782	1,063	4,517	957	3,892	877
	RA1-Corr[b]		***13.5***	**9,422**	**1,063**	**5,822**	**957**	**5,610**	**877**

[a] Data from Fig. 2.37

[b] ITSM values corrected based on Fig. X.38 for a ΔV_i of -1.9% for CBTM-EMU RA2 data (RA2-Corr) and for a ΔV_i of -2.6% for CBTM-FB RA1 (RA1-Corr)

the type of RA. Finally, for both CBTM-EMU and CBTM-FB, the RA source does not seem to show a significant difference in stiffness, even though the fragmentation test gives rise to a separate classification.

Relationship Between Stiffness and Temperature

Figure 2.40 shows the relationship between stiffness and temperature according to the test laboratory. Results show a significant variation in ITSM values, particularly

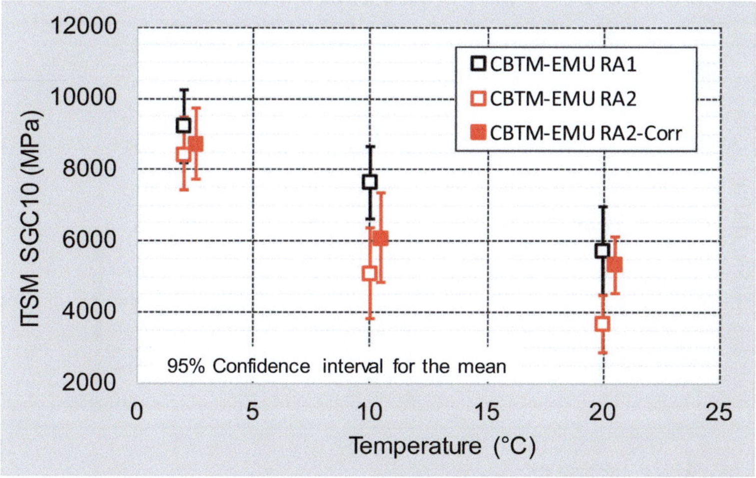

a)

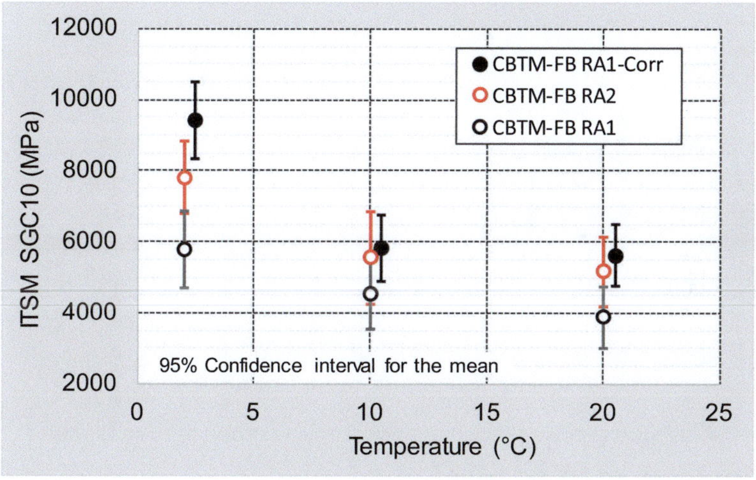

b)

Fig. 2.39 Interval plot of ITSM values corrected to account voids content differences (Data from SGC10 only): **a** CBTM-EMU, **b** CBTM-FB

for CBTM made with RA2. On the other hand, the results of LAB D include all the results obtained for the four types of compaction studied, whereas the results of LAB F were obtained from SGC 150 mm specimens. Nevertheless, the results of LAB A are in phase with those of LAB D for the tests on CBTM made with RA1.

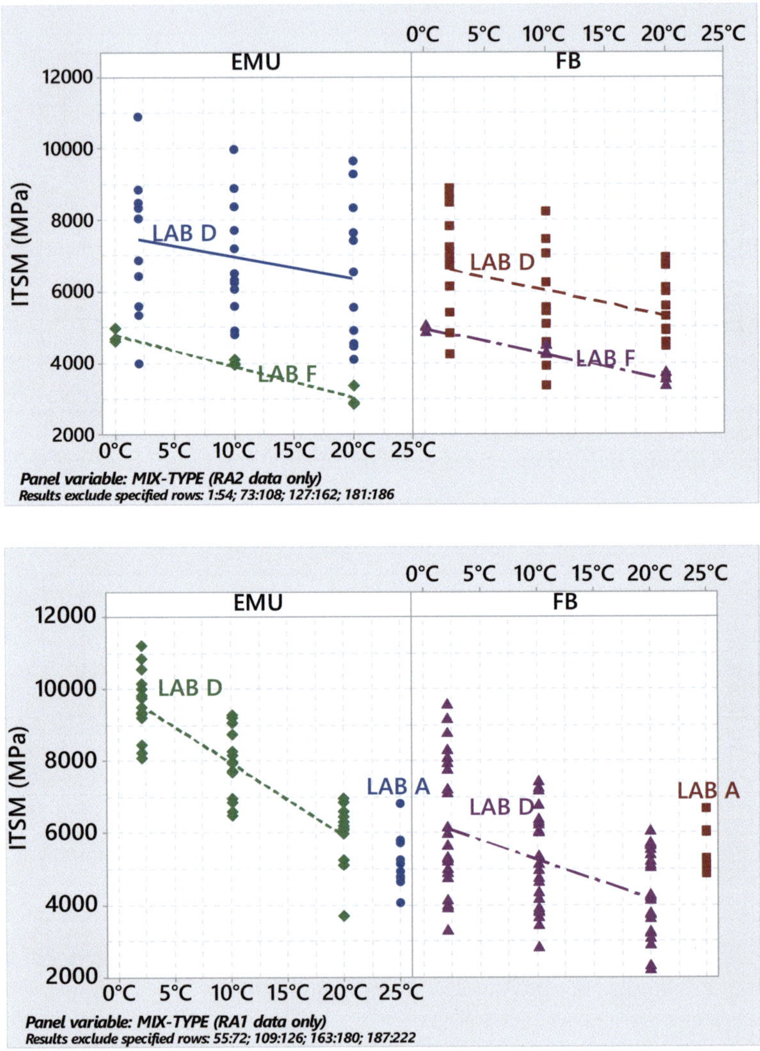

Fig. 2.40 Relationship between specimen's stiffness and the temperature: **a** data from RA2 only, **b** data from RA1 only

2.7.2.4 Conclusion

Although the results of this experimental campaign show that the ITS values of the CBTMs constructed with RA2 are generally higher, with a few exceptions, than those of the CBTMs constructed with RA1, the measured differences are not considered statistically significant. Moreover, ITS values on CBTM made of reconstituted granular fraction for RA1 and RA2 (same gradation and maximum size) [93] show an opposite tendency, i.e. the CBTM incorporating RA1 proved to be more resistant.

Furthermore, for both CBTM-EMU and CBTM-FB mixes, the RA source does not seem to show a significant difference in stiffness, even though the fragmentation test gives rise to a separate classification.

Ultimately, the mechanical performance (ITS and ITSM) of CBTM made from two RA sources, considered distinct according to their fragmentation resistance, are ultimately similar.

2.7.3 *Effect of Compaction on CBTM Mechanical Properties*

The laboratories involved in the inter-laboratory program used different compaction methods for the preparation of CBTM specimens. It is important to question whether the conclusions based on the measurement of mechanical properties are independent of the compaction method used to evaluate the mechanical performance of a CBTM.

2.7.3.1 ITS

Figure 2.41 shows the confidence intervals of the means of the indirect tensile strength (ITS) values of all the tests performed according to the compaction type. The results in Fig. 2.41 do not show a statistically significant difference in ITS values depending on the compaction method for the wet tested specimens but a significant difference for proctor compacted specimens tested in dry condition. Obviously, as pointed out earlier, the variation of void contents impacts the ITS values.

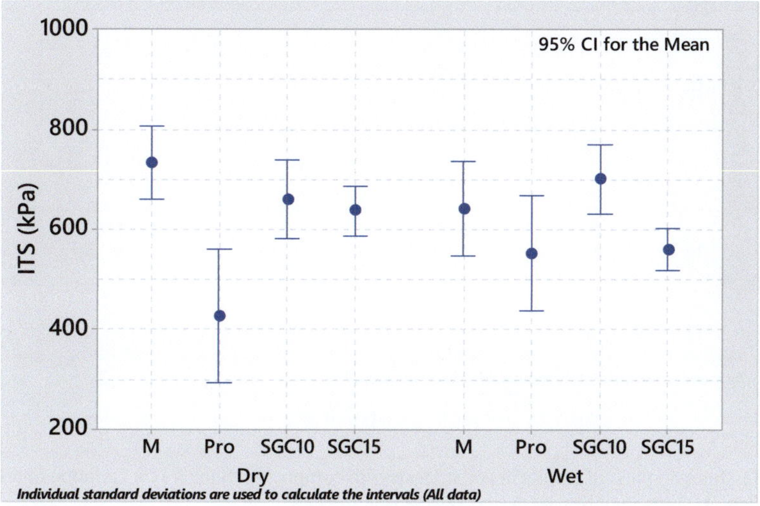

Fig. 2.41 Interval plots of ITS results in accordance with compaction type

Figure 2.42 shows the effect of compaction with reference to the type of CBTM. The results clearly show that only the ITS results of the CBTM-FB specimens compacted according to the Proctor method deviate from the ITS results obtained from the other compaction methods.

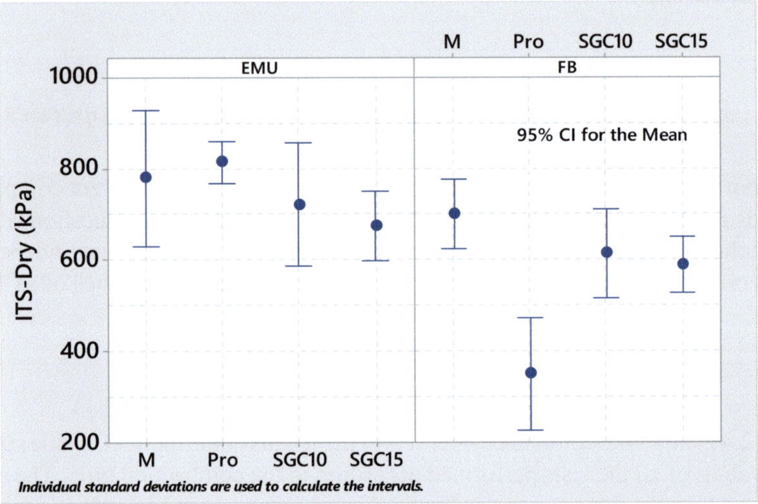

a)

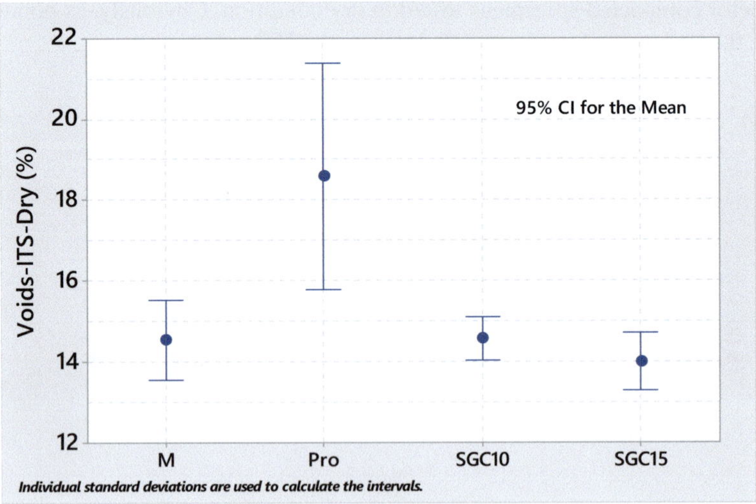

b)

Fig. 2.42 Interval plots of results in accordance with compaction type: **a** ITS variation in reference to mix type, **b** voids variation of dry compacted samples

However, when analyzing the average voids content and its variability according to the type of compaction (Fig. 2.42b), it can be seen that the average voids content in the specimens compacted using the Proctor method is very high. This means that the ITS values of these specimens are significantly lower and differ from those obtained from other compaction methods. On the other hand, the relative variability of the void contents for the Proctor specimens, established at 15.1%, is clearly greater than that calculated values for the specimens obtained from the three other compaction methods studied (M: 6.8%, SGC10: 3.7% and SGC15: 5.0%). In conclusion, it appears that regardless of the compaction method used to evaluate the mechanical performance of a CBTM in the laboratory, the ITS test is completely valid and comparable for the CBTM studied.

2.7.4 Effect of Curing Confinement

The method of curing the specimens remains a question mark regarding the measurement of the mechanical properties of a CBTM. We wish to use a curing method that has minimal variability and that is as representative as possible of site conditions. In this spirit, two curing methods were compared within the framework of this interlaboratory exchange program, namely a conventional cure: drying with free evaporation of all the surfaces of the cylindrical specimen (Free Surface Dry—FSD) and a restrictive cure: drying with free evaporation of the flat surfaces of the cylindrical specimen (Partially Surface Dry—PSD).

2.7.4.1 ITS

Figure 2.43 shows the histogram of ITS values from all data in accordance with the curing type. From the results presented in Fig. 2.43, the coefficients of variation according to the type of cure have been calculated and reported in Table 2.20. Likewise, the water resistance index (ITR) is given in Table 2.20 in correspondence with the type of cure considered. The variability according to the type of cure is distinct according to the test conditions: in wet conditions the variability of the ITS values for the PSD cure is clearly lower, but for the tests carried out in dry conditions the differences in variability between the PSD and FSD cure are practically negligible. We can underline the fact that the ITR values obtained testify to the good water resistance of all the CBTM studied.

Figure 2.44 shows the confidence intervals of the mean considering all the results. Figure 2.44a shows that the ITS values are statistically similar regardless of the curing method considered. On the other hand, by singling out the analysis to separate the effects of the cure according to the type of mixture (CBTM-EMU vs. CBTM-FB), the results in Fig. 2.44b) show statistically different ITS values according to the FSD type of cure, while for the PSD type of cure the results remain similar. Although these results cannot be linked to the actual values in the field, the FSD type of curing

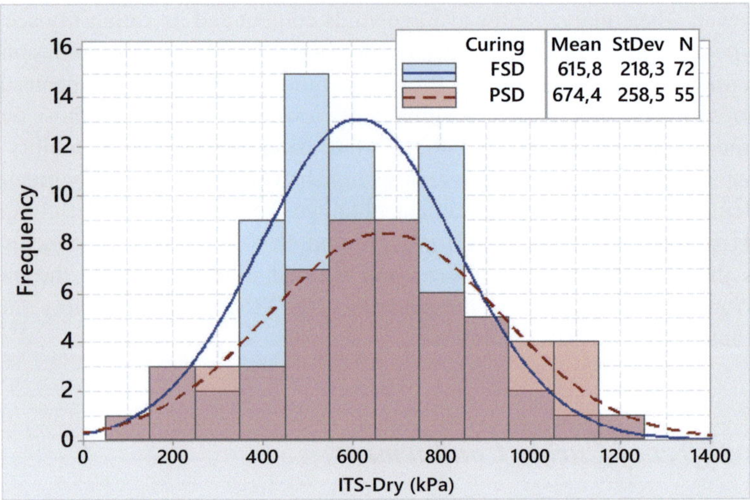

a)

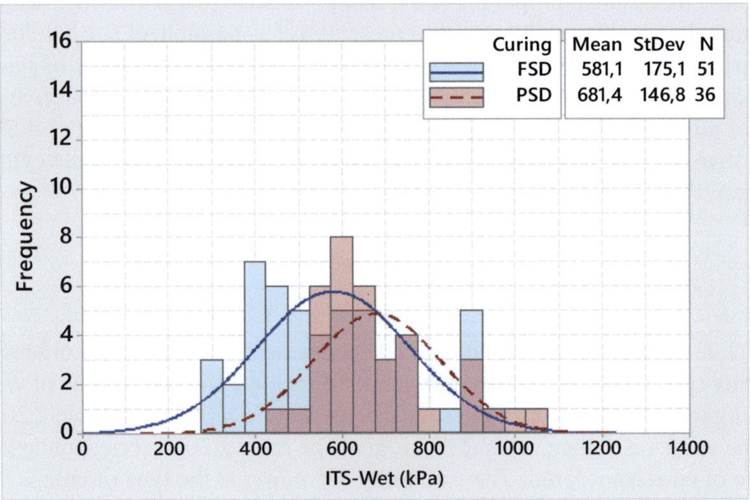

b)

Fig. 2.43 Histogram of ITS values in accordance with the curing type: **a** ITS-dry, **b** ITS-wet

Table 2.20 Summary of the results of Fig. 2.43

	(StDev/mean) * 100%		ITR
	Dry (%)	Wet (%)	(–)
FSD	35.4	30.1	0.94
PSD	38.3	21.5	1.01

predicts a difference in mechanical behavior between CBTM-FB and CBTM-EMU while the PSD type of curing predicts no difference. However, considering that the addition of bitumen and cement are the same for the two CBTM studied, it would be important to validate the relevance of one of these two curing methods based on field data.

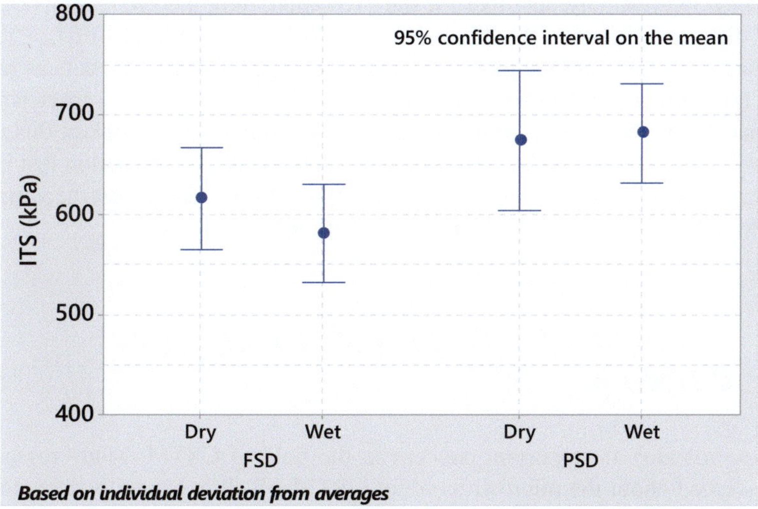

a)

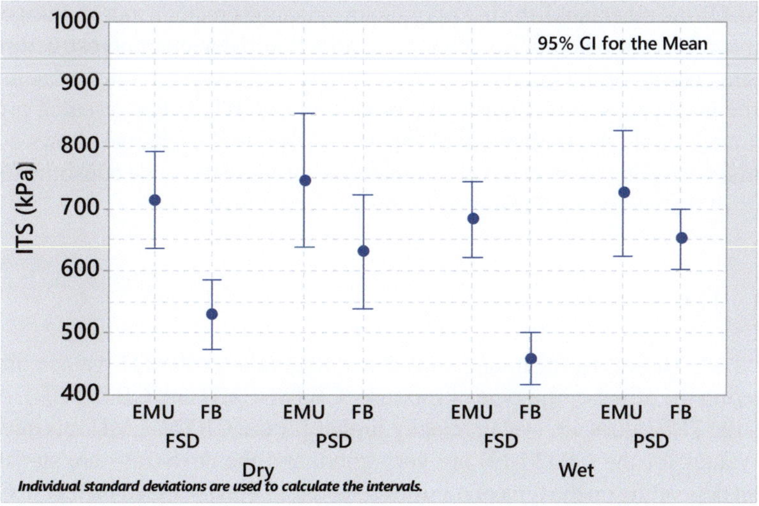

b)

Fig. 2.44 Interval plots of results in accordance to curing type: **a** effect of the testing condition, **b** effect of testing condition and mix type

2.7.4.2 ITSM

Figure 2.45a shows the confidence intervals of the mean ITSM considering all the results, and Fig. 2.45b shows the histogram of ITSM values from all data in accordance with the curing type. From the results presented in Fig. 2.45, it can be observed that the variability according to the curing time is somewhat similar for both types. Interestingly, just like for the ITS value, only the FSD results actually show a difference between the mix types (CBTM-EMU and CBTM-FB).

Figure 2.46 show interval plot of the means for the CBTM-EMU data as regards curing type. Figure 2.46a shows that the curing type only has an impact with the RA2, but the impact is not significant for RA1. Another way to look at this would be to say that the RAP source classified according to the fragmentation test is only visible with the PSD cure. Furthermore, Fig. 2.46b clearly shows that the variability of ITSM values is greater for CBTM-EMU with RA2.

2.7.5 *Effect of the Type of Mix: CBTM-EMU Versus CBTM-FB*

Mix type remains an important concern in the field of CBTM. Many researchers are concerned about the microstructure of CBTM-FB. It is generally accepted that the microstructure of a CBTM-FB is likely to present an aggregated microstructure where the added bitumen solidifies the granular skeleton at different points, lie spot welding. On the other hand, the use of bitumen emulsion creates a more homogeneous coating and has the potential to generate a more homogeneous microstructure with a uniform coating of the granular particles. These differences in microstructure can affect the mechanical performance and durability of CBTM. The research program orchestrated within the framework of this inter-laboratory exchange allows a closer comparison between these two types of materials because the same constituents were used by all the laboratories involved in this exchange program.

2.7.5.1 ITS

Figure 2.47a shows the mean and confidence intervals for the ITS values obtained considering the effect of the type of mixture (CBTM-EMU and CBTM-FB). At first glance, the ITS values are systematically higher for the CBTM-EMU mixture, even if ITS values for the CBTM-FB are very good, and the deviations are statistically different depending on the type of mixture considered. As discussed in Sect. 2.6.3, the compaction mode cannot explain this discrepancy. However, we can partly explain these differences by considering the variation in voids content of the specimens made with CBTM-EMU and CBTM-FB. Indeed, Fig. 2.47b shows that for all the specimens made and tested dry, the average voids content is clearly higher for the

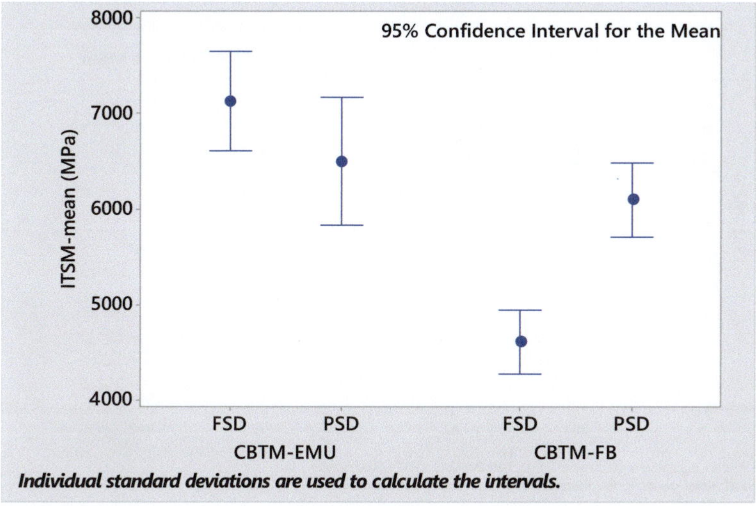

a)

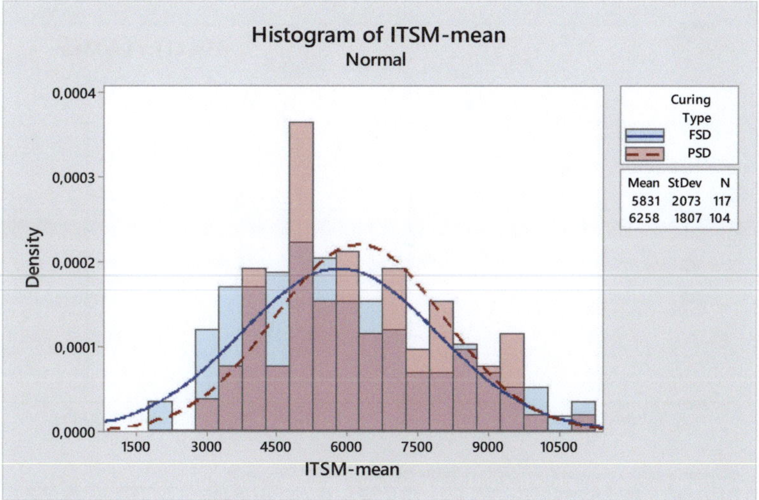

b)

Fig. 2.45 ITSM results: **a** mix and curing types effect, **b** variability in respect to curing type

specimens of the CBTM-FB mixture than for those of the CBTM-EMU mixture, i.e., respectively 15.8% (N = 54) and 13.8% (N = 45). Obviously, this difference has a direct impact on the ITS values.

Figure 2.48 shows the relationship between ITS values and voids contents for the two types of mixtures. The variability of the results is important and the impact of the void content is not negligible. A priori, it seems that the variation of the ITS value

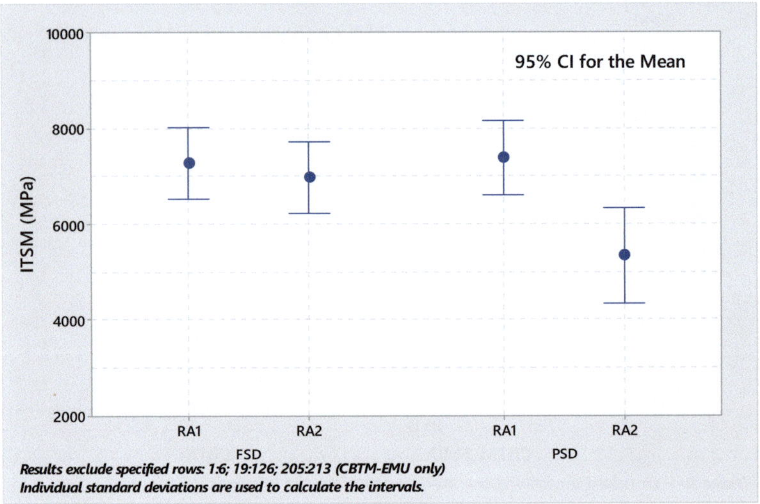

a)

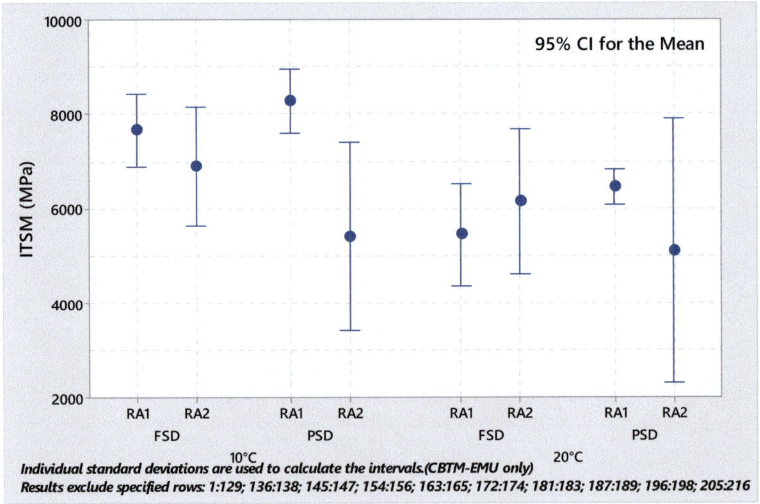

b)

Fig. 2.46 Interval plot of ITSM results in respect to the RA source as regards CBTM-EMU data only: **a** curing type effect, **b** temperature impact

in accordance to voids content is more marked for the CBTM-EMU mixture (slope of 91.8 compared to 56.2 for CBTM-FB).

Figure 2.49 records the results of specimens made only with the shear gyratory compactor (SGC). On the one hand, we believe that this compaction method is more appropriate to produce specimen's representative of the in-situ compaction mode and, on the other hand, this compaction mode allows a better control of voids

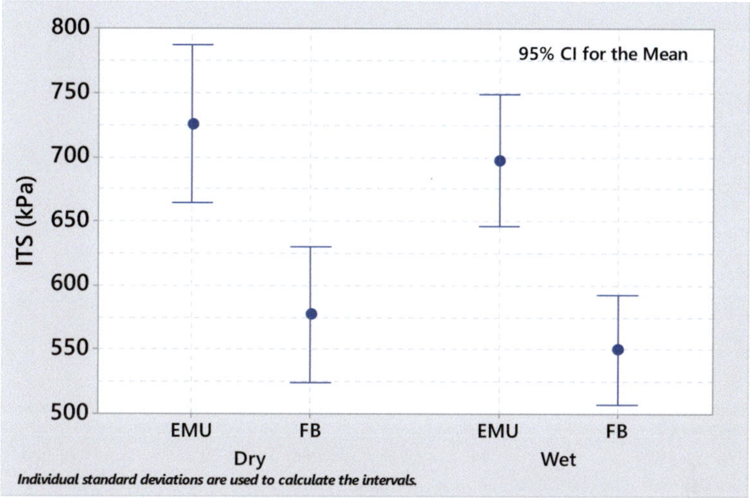

a)

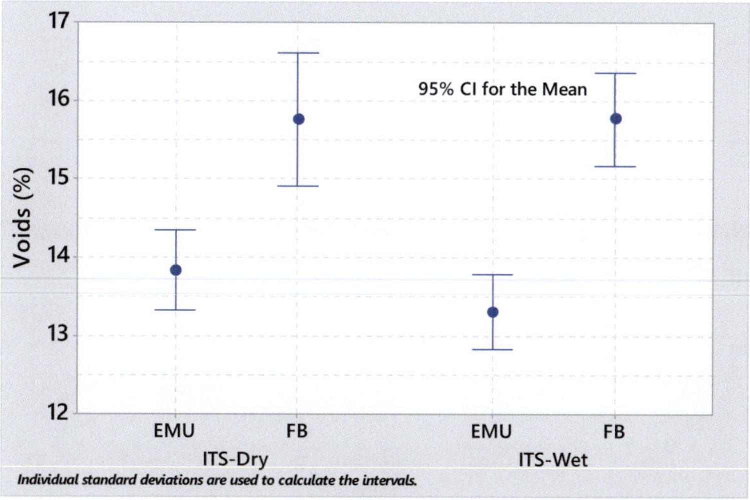

b)

Fig. 2.47 Interval plots of results in accordance to mix type: **a** ITS values, **b** voids content

independently of the type of mix. In this respect, Fig. 2.49a shows that the relative variation of voids in specimens compacted with SGC is relatively low, i.e., around 4% for all these results. On the other hand, the results show that the mean voids content is about 1% higher for the CBTM-FB specimens. The corresponding ITS values (Fig. 2.49b) here are quite comparative and without any deviation considered statistically representative. The ITS values for CBTM-EMU remain slightly higher. If we correct the average ITS value of the CBTM-EMU considering the difference

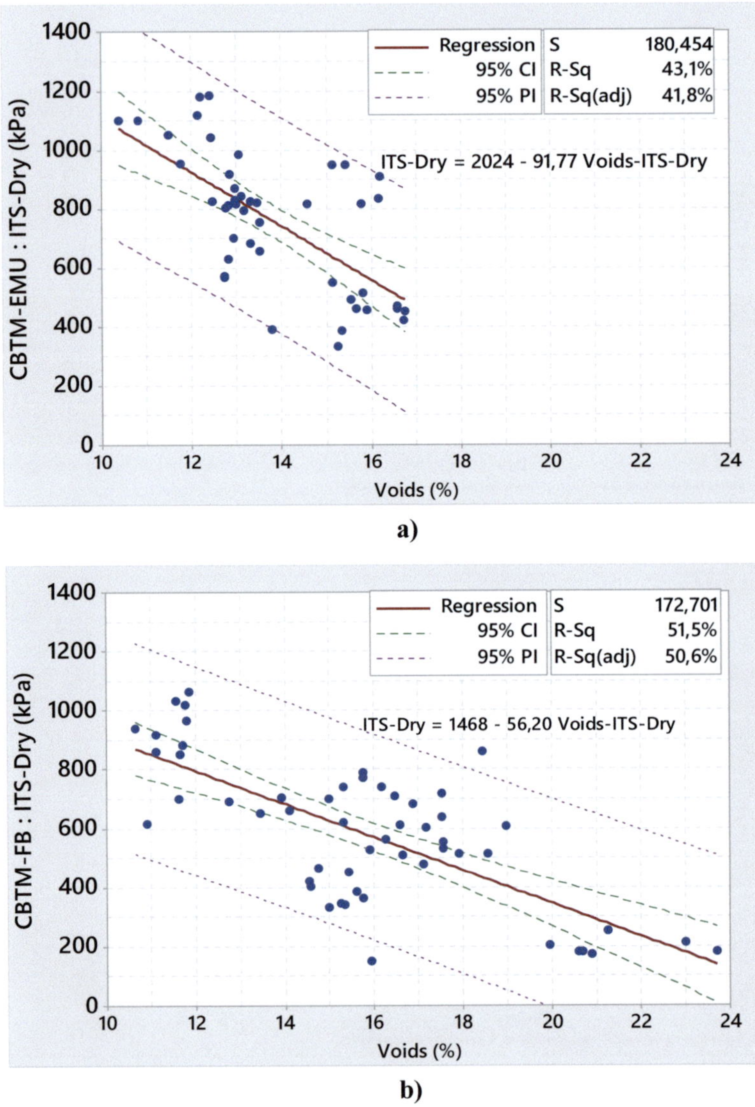

Fig. 2.48 Relationship between ITS values measured in dry state and the voids content: **a** CBTM-EMU, **b** CBTM-FB

in void content of about 1% while relying on the results of Fig. 2.48a), the corrected average ITS value of the tests carried out in the dry state goes from 692.6 to 604.8 kPa, which is equivalent to that obtained for the CBTM-FB (602.9 kPa). It can therefore be assumed that the mechanical performance evaluated from the ITS test shows no difference depending on the type of mix when using the same materials and at the same dosage.

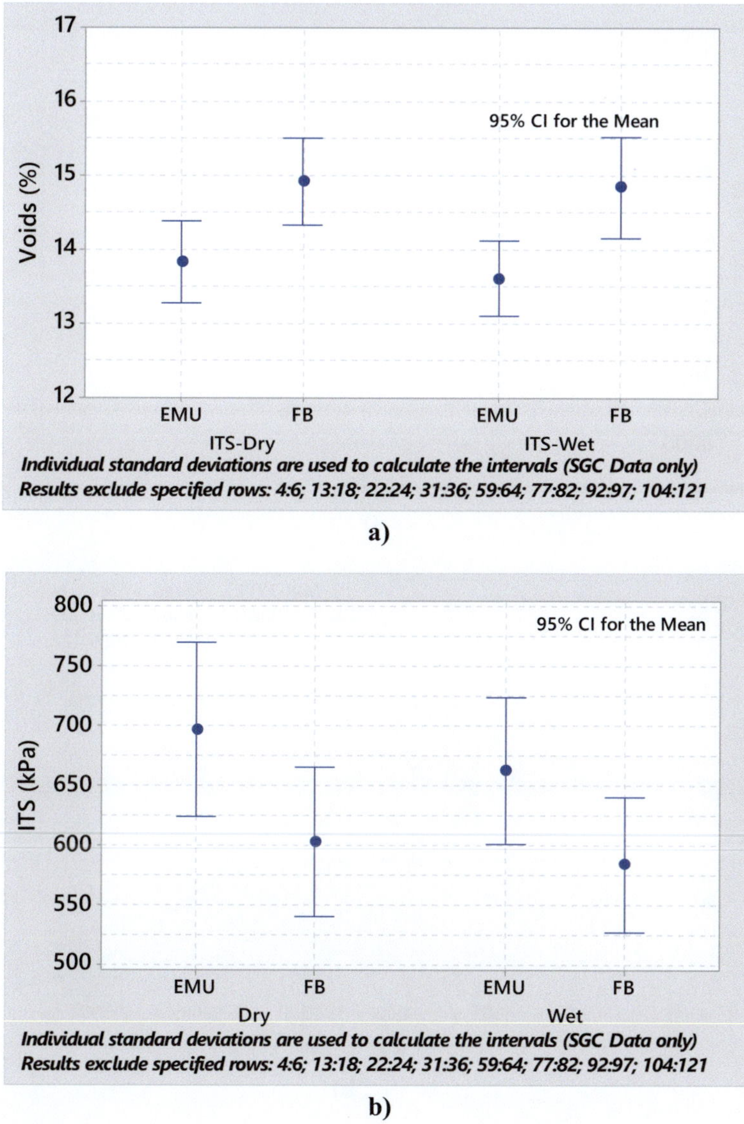

Fig. 2.49 Interval plots of results in accordance to mix type based on SGC data only: **a** voids content, **b** ITS values

2.7.5.2 ITSM

Figure 2.50a shows the confidence intervals of the mean ITSM considering all the results, and Fig. 2.50b shows the histogram of ITSM values from all data in accordance with the mix type.

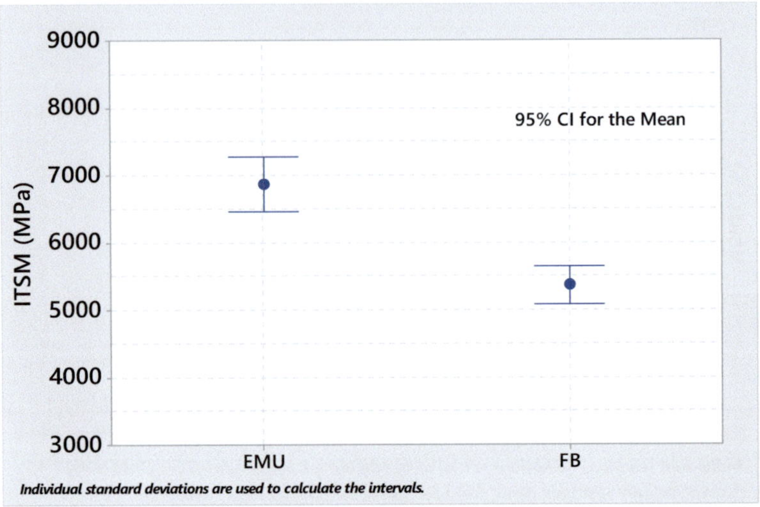

a)

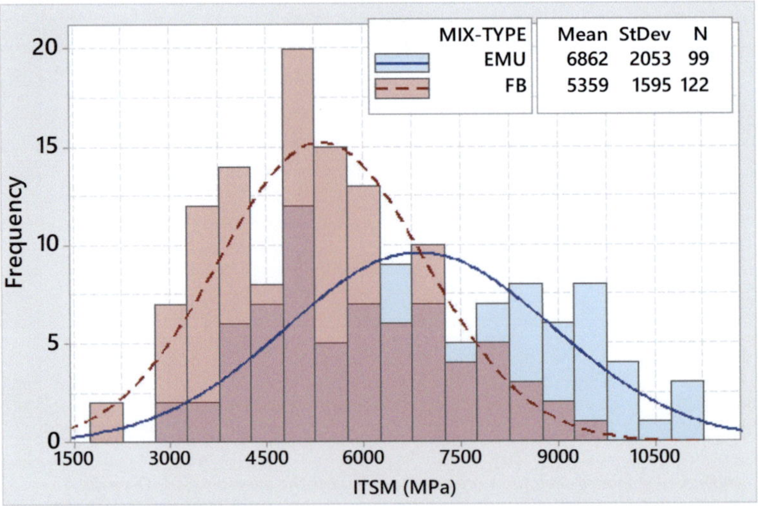

b)

Fig. 2.50 ITSM values in accordance to mix type. **a** Interval plots of results, **b** variability

Obviously, the difference in voids content has an impact on the ITSM values as shown in Fig. 2.36 (Sect. 2.7.4.2). Figure 2.51 shows the variation of the stiffness according to the voids content for the three main temperatures studied. Overall, the effect of voids is relatively similar for the two types of mixtures studied and it is slightly more pronounced at low temperatures.

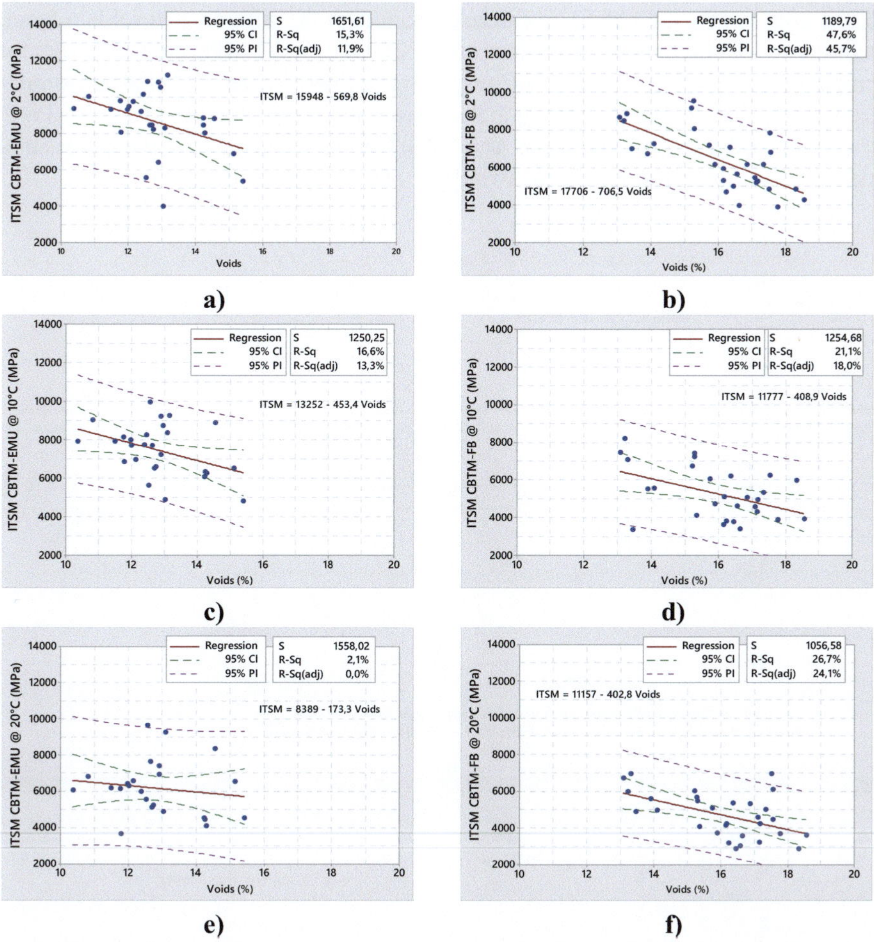

Fig. 2.51 Relationship between ITSM values and voids content in accordance to mix type: **a** CBTM-EMU (2 °C), **b** CBTM-FB (2 °C), **c** CBTM-EMU (10 °C), **d** CBTM-FB (10 °C), **e** CBTM-EMU (20 °C), **f** CBTM-FB (20 °C)

As with the analysis of the ITS results, we propose to refer to the results obtained from the SGC specimens to evaluate the impact of mixture type on ITSM values. The mean values and confidence intervals of the voids and ITSM values according to the type of mixture are reported in Fig. 2.52a, b. The relative variation in voids of the SGC specimens is small, less than 4.6%. On the other hand, the difference in voids content between the two types of mixtures studied is significant either for average values of 12.9% and 15.1% (difference of 2.2%) for CBTM-EMU and CBTM-FB respectively. As for the ITS results, Fig. 2.52b clearly shows that for similar void contents (SGC specimens), the ITSM values cannot be considered as statistically distinct and this, for all the considered temperatures. Referring to the slopes of the lines in Fig. 2.51, the

ITSM values of the CBTM-EMU can be corrected for the difference in voids content. In doing so the corrected ITSM values are 7,026 MPa, 5,671 MPa and 5,218 MPa for temperatures of 2 °C, 10 °C and 20C respectively. These values are relatively similar to those obtained for the CBTM-FB at the corresponding temperatures (6,963 MPa @ 2 °C, 5,287 MPa @ 2 °C and 5,218 @ 20 °C). It can be concluded that the measured stiffness's are quite similar for the two types of mixtures studied.

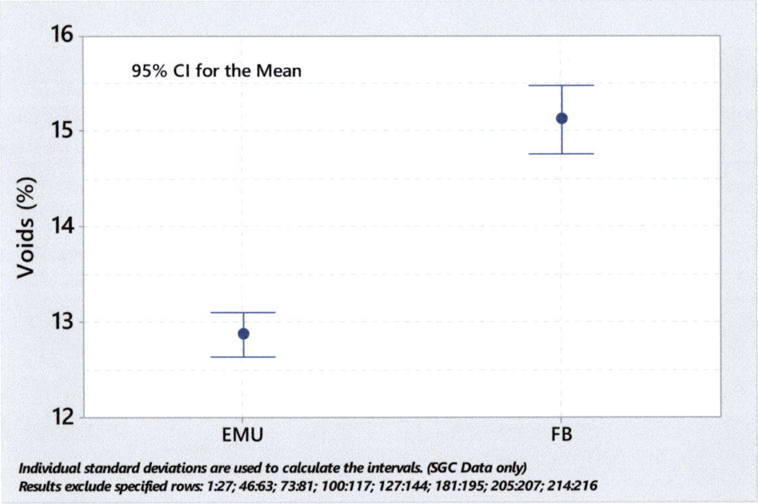

a)

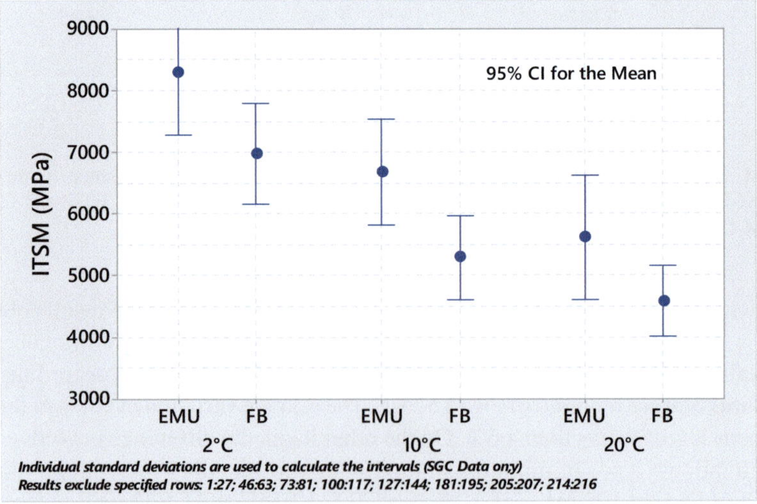

b)

Fig. 2.52 Interval plots of results in accordance to mix type based on SGC data only: **a** Voids content and **b** ITSM values

2.8 Conclusions

In this chapter, the overall activity of the RILEM TC 264-RAP Task Group 1 was presented. The use of the volumetric approach for CRM mix design has been proposed to better quantify the constituent proportions and control air voids of the CRM based on the same concept already used for hot-mix asphalt (HMA) mixtures. It is considered that this approach should be used for any CRM mix design procedure.

The use of cold recycling in pavement rehabilitation is becoming very common in some part of the world. Different materials can be used for cold recycling, and the properties and the performance of those mixtures change based on the choice of the components and the mix design. However, there is little information available and universally recognized on the impact of different factors used in the mix design, like the compaction method, the characteristics of the recycled material used, the curing process and the type of binder that is used to make the mix. It is known that the air voids content of CBTM or CRM in general are higher than hot mix asphalt, but there is limited information available on the expected air voids according to the mix design parameters and the methodology used to fabricate the specimens in laboratory.

The international ILTP including 12 laboratories from 10 countries was carried out to better understand the impact of different aspects related to CRM mix design. It was decided to concentrate TG1 efforts on the study of cement-bitumen treated materials (CBTM) in which bitumen emulsion (CBTM-EMU) or foamed bitumen (CBTM-FB) was used with a fixed amount (1.5%) of cement. The goals of this ILTP were to evaluate the ability of the fragmentation test to differentiate RAP source in terms of CBTM properties, to quantify the impact of the compaction method (SGC 100 and 150 mm, Marshall, Proctor) on CBTM properties, and finally to understand and quantify the impact of the curing protocol (free surface dry or partially surface dry) and the binder type on CBTM properties. It should be noted that all laboratories used exactly the same raw material (except water) and the same mix design following the TG1 test protocol to prepare, cure and test specimens.

Based on the research results the following conclusions separated into effects on the mix volumetric, air voids, and then effects on the mechanical properties can be drawn:

Volumetric:

- The volumetric approach used for HMA was extended to CRM taking into account specific peculiarities related to the presence of cement, water and bitumen in form of emulsion or foam and their evolutive conditions.
- The volumetric approach allows air voids, voids filled by liquid and void in mineral aggregate to be calculated and monitored; The results showed that the air voids were well-controlled with all the results being within 2% of the targeted value. However, there were differences in the air voids distribution depending on the type of mixes, CBTM-EMU and CBTM-FB, and between the different compaction methods (SGC 100 mm, SGC 150 mm, Marshall and Proctor). Overall, mixes made with foam had higher voids than mixes with emulsions.

As for the compaction, both SGC, 100 mm and 150 mm, and Marshall gave similar air voids and Proctor—lower air voids.

- The air voids content of the specimens compacted with CBTM-EMU was below the lower limit set while those compacted with CBTM-FB were above the upper limit set
- The higher air voids observed for the CBTM-FB mixes may be due to the differences in the fluid level between CBTM-FB and CBTM-EMU materials. For most labs, the mixes were produced in different batches that were mixed together before compaction, which normally took time. Regardless of the type of compaction recommended, compaction is always more difficult with the CBTM-FB mix.
- CBTM-EMU mixes were clearly below to the lower limit target for the testing program. An explanation is related to the fact that the hydrates generated by the hydration of Portland cement impact on the global balance of volumetric components and result ultimately in an apparent reduction of the volume of voids of the compacted CRM (V_m) over the time.
- As for compaction, results clearly showed that the compaction is more effective for the SGC 150 mm specimens rather than for SGC 100 mm and its relative variability is slightly lower (9.7% compared to 10.9% respectively). The SGC compaction method, with the use of the volumetric approach allowed a better control of the air voids content.

Mechanical properties:

- For both CBTM-EMU and CBTM-FB, the RA source did not lead to a significant difference in stiffness, even though the fragmentation test gives rise to a separate classification.
- Regardless of the compaction method used to evaluate the mechanical performance of a CBTM in the laboratory, the ITS test was completely valid and comparable for the CBTM studied.
- Although the results showed that the ITS values of the CBTMs constructed with RA2 were generally higher, with few exceptions, than those of the CBTMs constructed with RA1, the measured differences were not considered statistically significant. Moreover, ITS values on CBTM made of reconstituted granular fraction for RA1 and RA2 (same gradation and maximum size) showed an opposite tendency, i.e., the CBTM incorporating RA1 proved to be more resistant.
- The FSD type of curing predicts a difference in mechanical behavior (ITS and ITSM) between CBTM-FB and CBTM-EMU while the PSD type of curing predicts no difference. However, considering that the addition of bitumen and cement were the same for the two CBTM studied, it would be important to validate the relevance of one of these two curing methods based on field data.
- With respect to mix types, the ITS values were systematically higher for the CBTM-EMU mixture, even if the ITS values for the CBTM-FB are very good, and the deviations are statistically different depending on the type of mixture considered. Obviously, the difference in voids content remains a determining factor on the ITS values. By focusing the analysis on the specimens compacted with SGC (comparable void content), the ITS and ITSM values of the CBTM-EMU

always remain higher than those of the CBTM-FB but without being qualified as statistically different.

Acknowledgements The Task Group 1 of the RILEM Technical Committee on Asphalt Pavement Recycling (RAP) 264 would like to thank Nicola Frulli and Stefano Silvegni from CBR (Cooperativa Braccianti Riminese, Italy), NYNAS AB, the NCAT and Willemen Infra NV (Belgium) for their collaboration. The contribution of Andrea Baliello (University of Padova, Italy), Pieter Van Der Sypt (Willemen Infra NV, Belgium), Hilde Soenen (NYNAS AB, Belgium), Amir Rahmanbeiki and Marc-André Bérubé (LCMB-ÉTS), Francesco Preti and Bia Gouveia (University of Parma), Chibuike Ogbo (University of New Hampshire) and Ana Jiménez del Barco Carrión (University of Nottingham) to the experimental work is acknowledged.

References

1. Tebaldi, G., Dave, E.V., Cannone Falchetto, A., Hugener, M., Perraton, D., Grilli, A., Lo Presti, D., Pasetto, M., Loizos, A., Jenkins, K., Apeagyei, A., Grenfell, J., Bocci, M.: Recommendation of RILEM TC237-SIB: protocol for characterization of recycled asphalt (RA) materials for pavement applications. Mater. Struct. **51**(6), 142 (2018). https://doi.org/10.1617/s11527-018-1253-5
2. Partl, M.N., Porot, L., Di Benedetto, H., Canestrari, F., Marsac, P., Tebaldi, G.: Testing and characterization of sustainable innovative bituminous materials and systems. In: State-of-the-Art Report of the RILEM Technical Committee 237-SIB. Springer International Publishing (2018). https://doi.org/10.1007/978-3-319-71023-5
3. Grilli, A., Mignini, C., Graziani, A.: Field behaviour of cold-recycled asphalt mixtures for binder courses. In: Proceedings of the International Conference on Sustainable Materials, Systems and Structures (SMSS2019) New Generation of Construction Materials, pp. 124–131. RILEM Publications SARL (2019)
4. Ferrotti, G., Grilli, A., Mignini, C., Graziani, A.: Comparing the field and laboratory curing behaviour of cold recycled asphalt mixtures for binder courses. Materials **13**(21), 4697 (2020). https://doi.org/10.3390/ma13214697
5. Meocci, M., Grilli, A., La Torre, F., Bocci, M.: Evaluation of mechanical performance of cement–bitumen-treated materials through laboratory and in-situ testing. RMPD **18**(2), 376–389 (2017). https://doi.org/10.1080/14680629.2016.1213506
6. Bocci, M., Canestrari, F., Grilli, A., Pasquini, E., Lioi, D.: Recycling techniques and environmental issues relating to the widening of an high traffic volume Italian motorway. Int. J. Pavement Res. Technol. **3**(4), 171–177 (2010)
7. Bocci, M., Manganaro, A., Grilli, A.: Runway pavement reconstruction with full material recycling: the case of the airport of Treviso. Special Issue Adv. Mater. Res. **723**, 1044–1051 (2013). https://doi.org/10.4028/www.scientific.net/AMR.723.1044
8. Bocci, M., Grilli, A., Cardone, F., Ferrotti, G.: Full-depth reclamation for the rehabilitation of local roads: a case study. Int. J. Pavement Eng. **15**(3), 191–201 (2014). https://doi.org/10.1080/10298436.2012.657196
9. Tebaldi, G., Dave, E.V., Marsac, P., Muraya, P., Hugener, M., Pasetto, M., Graziani, A., Grilli, A., Bocci, M., Marradi, A., Wendling, L., Gaudefroy, V., Jenkins, K., Loizos, A., Canestrari, F.: Synthesis of standards and procedures for specimen preparation and in-field evaluation of cold-recycled asphalt mixtures. RMPD **15**(2), 272–299 (2014). https://doi.org/10.1080/14680629.2013.866707
10. Ziyani, L., Gaudefroy, V., Ferber, V., Deneele, D., Hammoum, F.: Chemical reactivity of mineral aggregates in aqueous solution: relationship with bitumen emulsion breaking. J. Mater. Sci. **49**(6), 2465–2476 (2014). https://doi.org/10.1007/s10853-013-7938-9

11. Bocci, M., Grilli, A., Cardone, F., Graziani, A.: A study on the mechanical behaviour of cement–bitumen treated materials. Constr. Build. Mater. **25**(2), 773–778 (2011). https://doi.org/10.1016/j.conbuildmat.2010.07.007
12. Grilli, A., Bocci, E., Graziani, A.: Influence of reclaimed asphalt content on the mechanical behaviour of cement-treated mixtures. RMPD **14**(3), 666–678 (2013). https://doi.org/10.1080/14680629.2013.794367
13. Sabita.: User guide for the design of asphalt mixes. Sabita: South Africa (2020). http://www.sabita.co.za/wp-content/uploads/2020/05/sabita-manual-24-third-revision-1.pdf. Accessed 24 July 2021
14. Collings, D., Grobler, J., Hughes, M., Jenkins, K., Jooste, F., Long, F., Thompson, H.: Technical guideline: Bitumen stabilized materials. A guideline for the design and construction of bitumen emulsion and foamed bitumen stabilised materials. TG2, 2nd edn. CSIR Built Environment, South Africa, Pretoria: Asphalt Academy (2009). https://www.asphaltacademy.co.za/wp-content/uploads/2016/09/TG2May09-1.pdf. Accessed 24 July 2021
15. Grilli, A., Graziani, A., Bocci, M.: Compactability and thermal sensitivity of cement–bitumen-treated materials. RMPD **13**(4), 599–617 (2012). https://doi.org/10.1080/14680629.2012.742624
16. Grilli, A., Cardone, F., Bocci, E.: Mechanical behaviour of cement-bitumen treated materials containing different amounts of reclaimed asphalt. Eur. J. Environ. Civ. Eng. **22**(7), 836–851 (2018). https://doi.org/10.1080/19648189.2016.1219972
17. Xu, O., Wang, Z., Wang, R.: Effects of aggregate gradations and binder contents on engineering properties of cement emulsified asphalt mixtures. Constr. Build. Mater. **135**, 632–640 (2017). https://doi.org/10.1016/j.conbuildmat.2016.12.095
18. Perraton, D., Tebaldi, G., Dave, E., Bilodeau, F., Giacomello, G., Grilli, A., Graziani, A., Bocci, M., Grenfell, J., Muraya, P., Pasetto, M., Kuna, K., Apeagyei, A., Lo Presti, D., Airey, G., Jenkins, K., Hajj, E., Hugener, M, Marsac, P.: Tests campaign analysis to evaluate the capability of fragmentation test to characterize recycled asphalt pavement (RAP) material. In: Canestrari, F., Partl, M.N. (eds.) 8th RILEM International Symposium on Testing and Characterization of Sustainable and Innovative Bituminous Materials, pp. 965–976. Dordrecht, Springer Netherlands (2016). https://doi.org/10.1007/978-94-017-7342-3_77
19. Tebaldi, G., Dave, E., Cannone Falchetto, A., Hugener, M., Perraton, D., Grilli, A., Lo Presti, D., Pasetto, M., Loizos, A., Jenkins, K., Apeagyei, A., Grenfell, J., Bocci, M.: Recommendation of RILEM TC237-SIB on fragmentation test for recycled asphalt. Mater. Struct. **52**(4), 82 (2019). https://doi.org/10.1617/s11527-019-1365-6
20. Cui, D., Pang, J.: The effect of pH on the properties of a cationic bitumen emulsifier. Tenside Surfactants Deterg. **54**, 5 (2017). https://doi.org/10.3139/113.110520accessedon24-07-2021
21. James, A.: Overview of asphalt emulsion, Emulsion Asphalt Technology, Transportation Research Circular, Number E-C012 (2006)
22. Jenkins, K.J., van de Ven, M.F.C., de Groot, J.L.A.: characterisation of foamed bitumen. In: 7th Conference on Asphalt Pavements for Southern Africa (1999)
23. Jenkins, K.J.: Mix design considerations for cold and half-warm bituminous mixes with emphasis on foamed bitumen. Ph.D. thesis, University of Stellenbosch, South Africa (2000). https://www.loudoninternational.co.za/wp-content/uploads/2020/09/Jenkins-PhD-Cold-and-Halfwarm-Foamed-bitumen-.pdf. Accessed 24 July 2021
24. EN_933-1.: Tests for geometrical properties of aggregates—Part 1: determination of particle size distribution—Sieving method. European Committee for Standardization, Brussels (2012)
25. EN_1097-7.: Tests for mechanical and physical properties of aggregates—Part 7: Determination of the particle density of filler—Pyknometer method. European Committee for Standardization, Brussels (2008)
26. EN_933-9+A1.: Tests for geometrical properties of aggregates—Part 9: assessment of fines—Methylene blue test. European Committee for Standardization, Brussels (2013)
27. EN_196-6.: Methods of testing cement—Part 6: determination of fineness. European Committee for Standardization, Brussels (2019)

28. EN_1097-4.: Tests for mechanical and physical properties of aggregates—Part 4: determination of the voids of dry compacted filler. European Committee for Standardization, Brussels (2008)
29. EN_13179-1.: Tests for filler aggregate used in bituminous mixtures—Part 1: delta ring and ball test. European Committee for Standardization, Brussels (2013)
30. EN_196-21.: Methods of testing cement—Determination of the chloride, carbon dioxide and alkali content of cement. European Committee for Standardization, Brussels (1991)
31. Yan, J., Leng, Z., Li, F., Zhu, H., Bao, S.: Early-age strength and long-term performance of asphalt emulsion cold recycled mixes with various cement contents. Constr. Build. Mater. **137**, 153–159 (2017). https://doi.org/10.1016/j.conbuildmat.2017.01.114
32. EN_196-1.: Methods of testing cement—Part 1: determination of strength. European Committee for Standardization, Brussels (2016)
33. EN_196-3.: Methods of testing cement—Part 3: determination of setting times and soundness. European Committee for Standardization, Brussels (2016)
34. Niazi, Y., Jalili, M.: Effect of Portland cement and lime additives on properties of cold in-place recycled mixtures with asphalt emulsion. Constr. Build. Mater. **23**(3), 1338–1343 (2009). https://doi.org/10.1016/j.conbuildmat.2008.07.020
35. EN_549.: Rubber materials for seals and diaphragms for gas appliances and gas equipment. European Committee for Standardization, Brussels (2019)
36. EN_459-2.: Building lime—Part 2: test methods. European Committee for Standardization, Brussels (2021)
37. Bonfim, V.: Cold milling of asphalt pavements. 1st ed. EXCEÇAO (2011)
38. AASHTO_MP31-17.: Standard method of test for materials for cold recycled mixtures with emulsified asphalt, AASHTO provisional standards. American Association of State Highway and Transportation Officials, Washington, D.C. (2017)
39. NQ2560-114.: Travaux de génie civil—Granulats. BNQ. Bureau de normalisation du Québec (2002)
40. Raschia, S., Graziani, A., Carter, A., Perraton, D.: Laboratory mechanical characterisation of cold recycled mixtures produced with different RAP sources. RMPD **20**(sup1), S233–S246 (2019). https://doi.org/10.1080/14680629.2019.1588775
41. Ghavibazzo, A., Attia, M., Soltis, P., Ajideh, H.: Effect of gradation and aged binder content of reclaimed asphalt pavement (RAP) on properties of cold-recycled asphalt mix. In: International Conference on Highway Pavements and Airfield Technology Proceedings, Philiadelphia, USA (2018). https://doi.org/10.1061/9780784480946.008
42. Gandi, A., Cardenas, A., Sow, D., Carter, A., Perraton, D.: Study of the impact of the compaction and curing temperature on the behavior of cold bituminous recycled materials. J. Traffic Transp. Eng. **6**(4), 349–358 (2019). https://doi.org/10.1016/j.jtte.2019.03.002
43. Pasetto, M., Baliello, A., Giacomello, G., Pasquini, E.: Cold recycling of reclaimed asphalt: analysis of alternative procedures. In: Bituminous Mixtures and Pavements VII—Proceedings, 7th International Conference on Bituminous Mixtures and Pavements (7th ICONFBMP), Thessaloniki, Greece (2019)
44. Gandi, A.: Laboratory characterization of bitumen treated full depth reclamation materials. Ph.D. thesis, Université du Québec, Montreal, Canada (2018). https://espace.etsmtl.ca/id/eprint/2140/1/GANDI_Apparao.pdf. Accessed 24 July 2021
45. Raschia, S., Perraton, D., Graziani, A., Carter, A.: Influence of low production temperatures on compactability and mechanical properties of cold recycled mixtures. Constr. Build. Mater. **232**, 117169 (2020). https://doi.org/10.1016/j.conbuildmat.2019.117169
46. Tebaldi, G., Marsac, P., Dave, E., Muraya, P., Hugener, M., Graziani, A., Grilli, A., Marradi, A., Gaudefroy, V., Jenkins, K., Loizos, A., Bocci, M., Wendling, L.: Synthesis of specimen preparation and curing processes for cold recycled asphalt mixes. In: 2nd International Symposium on Asphalt Pavement and Environment, ISAP, Fortaleza (2012)
47. MNDOT.: Special provisions—SP2005Book. June 30, 2006. Last revision 4/16/08 (2008). http://www.dot.state.mn.us/materials/bituminousdocs/Pavement%20Maintenance/SP2005-130-2331-cir-with-train.pdf. Accessed 26 Oct. 2021

48. Cardone, F., Grilli, A., Bocci, M., Graziani, A.: Curing and temperature sensitivity of cement–bitumen treated materials. Int. J. Pavement Eng. **16**(10), 868–880 (2015). https://doi.org/10.1080/10298436.2014.966710
49. Xiao, F., Yao, S., Wang, J., Li, X., Amirkhanian, S.: A literature review on cold recycling technology of asphalt pavement. Constr. Build. Mater. **180**, 579–604 (2018). https://doi.org/10.1016/j.conbuildmat.2018.06.006
50. Grilli, A., Graziani, A., Bocci, E., Bocci, M.: Volumetric properties and influence of water content on the compactability of cold recycled mixtures. Mater. Struct. **49**(10), 4349–4362 (2016). https://doi.org/10.1617/s11527-016-0792-x
51. EN_12697-5.: Bituminous mixtures—Test methods—Part 5: determination of the maximum density. European Committee for Standardization, Brussels (2019)
52. EN_12697-6.: Bituminous mixtures. Test methods. Determination of bulk density of bituminous specimens. European Committee for Standardization, Brussels (2020)
53. Dołżycki, B., Grilli, A., Balzi, A., Jaczewski, M., Szydłowski, C.: Binder courses using cold recycled mixtures: a novel concept in cold recycling. In: Proceedings of the 6th International Conference on Road and Rail Infrastructure (CETRA 2020), Pula, Croatia (2020). Accessed 20–21 May 2021
54. AASLP Azienda Autonoma di Stato per i Lavori Pubblici [in Italian] (2020). https://www.aaslp.sm/
55. Graziani, A., Godenzoni, C., Cardone, F., Bocci, M.: Effect of curing on the physical and mechanical properties of cold-recycled bituminous mixtures. Mater. Des. **95**, 358–369 (2016). https://doi.org/10.1016/j.matdes.2016.01.094
56. Graziani, A., Iafelice, C., Raschia, S., Perraton, D., Carter, A.: A procedure for characterizing the curing process of cold recycled bitumen emulsion mixtures. Constr. Build. Mater. **173**, 754–762 (2018). https://doi.org/10.1016/j.conbuildmat.2018.04.091
57. ASTM_D1556/M1556-15e1.: Standard test method for density and unit weight of soil in place by sand-cone method. American Society for Testing and Materials, West Conshohocken, PA, USA (2015). https://doi.org/10.1520/D1556_D1556M-15E01
58. EN_12697-23.: Bituminous mixtures—Test methods—Part 23: determination of the indirect tensile strength of bituminous specimens. European Committee for Standardization, Brussels (2017)
59. ASTM_D6931-17.: Standard test method for indirect tensile (IDT) strength of asphalt mixtures. American Society for Testing and Materials, West Conshohocken, PA, USA (2017). https://doi.org/10.1520/D6931-17
60. EN_12697-12.: Bituminous mixtures—Test methods—Part 12: determination of the water sensitivity of bituminous specimens. European Committee for Standardization, Brussels (2018)
61. ASTM_D4867/D4867M-09.: Standard test method for effect of moisture on asphalt concrete paving mixtures. American Society for Testing and Materials, West Conshohocken, PA, USA (2014). https://doi.org/10.1520/D4867_D4867M-09R14
62. EN_12697-26.: Bituminous mixtures—Test methods—Part 26: stiffness. European Committee for Standardization, Brussels (2018)
63. ASTM_D6307-19.: Standard test method for asphalt content of asphalt mixture by ignition method. American Society for Testing and Materials, West Conshohocken, PA, USA (2019). https://doi.org/10.1520/D6307-19
64. ASTM_D448-03.: Standard classification for sizes of aggregate for road and bridge construction. American Society for Testing and Materials, West Conshohocken, PA, USA (2003). https://doi.org/10.1520/D0448-03
65. ASTM_C127-15.: Standard test method for relative density (specific gravity) and absorption of coarse aggregate. American Society for Testing and Materials, West Conshohocken, PA, USA (2015). https://doi.org/10.1520/C0127-15
66. EN_1097-6.: Tests for mechanical and physical properties of aggregates—Part 6: determination of particle density and water absorption. European Committee for Standardization, Brussels (2013)

67. ASTM_D1557-12.: Standard test methods for laboratory compaction characteristics of soil using modified effort (56,000 ft-lbf/ft^3 (2,700 kN m/m^3. American Society for Testing and Materials)), West Conshohocken, PA, USA (2021). https://doi.org/10.1520/D1557-12R21
68. EN_932-2.: Tests for general properties of aggregates—Part 2: methods for reducing laboratory samples. European Committee for Standardization, Brussels (1999)
69. ASTM_C702/C702-18.: Standard practice for reducing samples of aggregate to testing size. American Society for Testing and Materials, West Conshohocken, PA, USA (2018). https://doi.org/10.1520/C0702_C0702M-18
70. ASTM_D2397/D2397M-20.: Standard specification for cationic emulsified asphalt. American Society for Testing and Materials, West Conshohocken, PA, USA (2020). https://doi.org/10.1520/D2397_D2397M-20
71. EN_1430.: Bitumen and bituminous binders—Determination of particle polarity of bituminous emulsions: European Committee for Standardization, Brussels (2009)
72. EN_13075-1.: Bitumen and bituminous binders—Determination of breaking behaviour—Part 1: determination of breaking value of cationic bituminous emulsions, mineral filler method. European Committee for Standardization, Brussels (2016)
73. EN_12848.: Bitumen and bituminous binders—Determination of mixing stability with cement of bituminous emulsions. European Committee for Standardization, Brussels (2009)
74. EN_1428.: Bitumen and bituminous binders—Determination of water content in bituminous emulsions—Azeotropic distillation method. European Committee for Standardization, Brussels (2012)
75. EN_1431.: Bitumen and bituminous binders—Determination of residual binder and oil distillate from bitumen emulsions by distillation. European Committee for Standardization, Brussels (2018)
76. EN_12846.: Bitumen and bituminous binders—Determination of efflux time of bitumen emulsions by the efflux viscometer. European Committee for Standardization, Brussels (2002)
77. EN_12846-1.: Bitumen and bituminous binders—Determination of efflux time by the efflux viscometer—Part 1: bituminous emulsions. European Committee for Standardization, Brussels (2011)
78. EN_1429.: Bitumen and bituminous binders—Determination of residue on sieving of bituminous emulsions, and determination of storage stability by sieving. European Committee for Standardization, Brussels (2013)
79. EN_12847.: Bitumen and bituminous binders—Determination of settling tendency of bituminous emulsions. European Committee for Standardization, Brussels (2009)
80. EN_13614.: Bitumen and bituminous binders—Determination of adhesivity of bituminous emulsions by water immersion test. European Committee for Standardization, Brussels (2021)
81. EN_1426.: Bitumen and bituminous binders—Determination of needle penetration. European Committee for Standardization, Brussels (2015)
82. EN_1427.: Bitumen and bituminous binders—Determination of the softening point—Ring and Ball method. European Committee for Standardization, Brussels (2015)
83. CSA_A3000-18.: Cementitious materials compendium. Canadian standards association (2018)
84. ASTM_C109/C109M-20b.: Standard test method for compressive strength of hydraulic cement mortars (using 2-in. or [50 mm] cube specimens. American Society for Testing and Materials, West Conshohocken, PA, USA (2020). https://doi.org/10.1520/C0109_C0109M-20B
85. ASTM_C188.: Standard test method for density of hydraulic cement. American Society for Testing and Materials, West Conshohocken, PA, USA (2017). https://doi.org/10.1520/C0188-17
86. ASTM_C204-18e1.: Standard test Methods for fineness of hydraulic cement byair-permeability apparatus. American Society for Testing and Materials, West Conshohocken, PA, USA (2018). https://doi.org/10.1520/C0204-18E01
87. ASTM_C430.: Standard test method for fineness of hydraulic cement by the 45 μm (No. 325) sieve. American Society for Testing and Materials, West Conshohocken, PA, USA (2017). https://doi.org/10.1520/C0430-17

88. CSA_A3004-B2.: Test method for time of set of hydraulic cement paste—Vicat needle. Canadian standards association (2018)
89. CSA_A3004-C2.: Test method for determination of compressive strengths. Canadian standards association (2018)
90. LC_26-002.: Méthode de formulation à froid des matériaux recyclés stabilisés à l'émulsion
91. Grilli, A., Raschia, S., Perraton, D., Carter, A., Rahmanbeiki, A., Kara De Maeijer, P., Lo Presti, D., Airey, G., Ogbo, C., Dave, E.V., Tebaldi, G.: Experimental investigation on water loss and stiffness of CBTM using different RA sources. In: Di Benedetto, H., Mangiafico, S., Chailleux, E., Sauzéat, C., Baaj, H., Tebaldi, G. (eds.) Proceedings of RILEM International Symposium of Bituminous Materials, Lyon, France (2020). https://doi.org/10.1007/978-3-030-46455-4_2
92. Pasetto, M., Pasquini, E., Baliello, A., Raschia, S., Rahmanbeiki, A., Carter, A., Perraton, D., Preti, F., Chagas Silva Gouveia, B., Tebaldi, G., Grilli, A., Dave, E.: Influence of curing on the mechanical properties of cement-bitumen treated materials using foamed bitumen: An inter-laboratory test program. In: Proceedings of the 9th International Conference on Maintenance and Rehabilitation of Pavements. Zurich (2020)
93. Raschia, S., Moghaddam Taher, B., Perraton, D., Baaj, H., Carter, A., Graziani, A.: Effect of RAP source on compactability and behaviour of cold-recycled mixtures in the small strain domain. J. Mater. Civ. Eng. **33**(4), 04021030 (2021). https://doi.org/10.1061/(ASCE)MT.1943-5533.0003614

Chapter 3
Evaluation of Warm Mix Asphalts with High Reclaimed Asphalt Content

Paul Marsac, Edoardo Bocci, Fabrizio Cardone, Augusto Cannone Falchetto, Di Wang, Xavier Carbonneau, Martins Zaumanis, Alan Carter, M. Carmen Rubio-Gámez, Miguel Del Sol Sánchez, Eshan V. Dave, and Gabriele Tebaldi

P. Marsac
Université Gustave Eiffel, Bouguenais, France
e-mail: paul.marsac@ifsttar.fr

E. Bocci
eCampus University, Novedrate, Italy
e-mail: edoardo.bocci@uniecampus.it

F. Cardone
Department of Civil and Building Engineering and Architecture, Università Politecnica Delle Marche, Ancona, Italy
e-mail: f.cardone@staff.univpm.it

A. C. Falchetto
Dipartimento di Ingegneria Civile, Edile e Ambientale, Universita degli studi di Padova, Padova, Italy
e-mail: augusto.cannone@unipd.it

D. Wang
Department of Civil Engineering, University of Ottawa, Ottawa, Canada
e-mail: dwang6@uOttawa.ca

X. Carbonneau
Campus Scientifique Et Technique (CST) COLAS, Magny-Les-Hameaux, France
e-mail: carbonneau@campus.colas.fr

M. Zaumanis
Empa—Materials Science and Technology, Dübendorf, Switzerland
e-mail: martins.zaumanis@empa.ch

A. Carter
Department of Construction Engineering, École de Technologie Supérieure (LCMB-ETS), Montreal, QC, Canada
e-mail: alan.carter@etsmtl.ca

G. Tebaldi and E. Dave (eds.), *Recycling Asphalt Pavement*, RILEM State-of-the-Art Reports 44, https://doi.org/10.1007/978-3-031-94231-0_3

Keywords Warm mix asphalt · Reclaimed asphalt · Mechanical performance · Sustainability · Interlaboratory testing · Mix design · Binder characterization

3.1 Motivation

The next generation of asphalt roads need to be sustainable to comply with the current demands of society. This trend started some years ago, trying to provide asphalt mixtures and manufacturing techniques that are more environmentally friendly. Under this goal, through these technologies, it will be possible to preserve natural resources by minimizing their consumption, eliminating waste generation, and reducing the atmospheric emissions and energy consumption associated with the manufacturing and laying asphalt materials.

Among the different technologies that can produce more sustainable asphalt mixtures, using reclaimed asphalt (RA) as a substitute for virgin aggregates in combination with a decrease in manufacturing temperature is considered the most promising option. Nonetheless, the materials' service life is also a key factor to be considered because the most durable material might be the most sustainable.

This is the current focus for the Task Group 2 of the RILEM Technical Committee 264-RAP on non-cold recycling, which has identified the need for research on asphalt mixtures manufactured at low temperatures, incorporating RA among its components, to obtain a performance similar to conventional asphalt mixtures (HMA). Literature shows that with a reduction of 20–30 °C in temperature, it is possible to obtain warm mix asphalt (WMA) with a similar performance to HMA. However, various issues associated with the durability of the combination of WMA + RA have also been reported.

Therefore, the motivation of this Task Group (TG) was focused on studying how the combination of WMA + RA might affect the mechanical properties of a conventional HMA. For this purpose, an international inter-laboratory testing program was launched in six different laboratories, which together developed a methodology derived from challenges that a road materials designer commonly faces: reduction in temperature with RA while preserving similar mechanical properties. The tested

M. Carmen Rubio-Gámez · M. Del Sol Sánchez
Laboratory of Construction Engineering, University of Granada, Granada, Spain
e-mail: mcrubio@ugr.es

E. V. Dave (✉)
Department of Civil and Environmental Engineering, University of New Hampshire, Durham, NH, USA
e-mail: eshan.dave@unh.edu

G. Tebaldi (✉)
Department of Engineering and Architecture, University of Parma, Parma, Italy
e-mail: gabriele.tebaldi@unipr.it

materials were characterized according to the internal testing protocol of each participant laboratory. This consisted of two variations of the same type of mixture (similar grading curve, binder content, etc.): a conventional Hot Mix Asphalt (HMA) without RA taken as reference and a WMA with 40% RA. The relative performances (workability, water sensitivity, permanent deformations, bearing capacity and fatigue) of all the WMA with 40% RA were then compared to identify common trends from the different characterization methods and possible performance issues derived from the WMA + RA combinations.

3.2 Background

3.2.1 WMA Technologies: Additives and Foaming

Over the years, Warm Mix Asphalt (WMA) has received significant attention as a technology capable of orienting the design and construction of asphalt pavement toward a more sustainable process. WMA benefits are obtained by lowering the production temperature of the asphalt mixture compared to the conventional Hot Mix Asphalt (HMA). Mixing and production plants can achieve this target using different technologies, mainly consisting of additives or foaming technology. The selection of the specific technical solution depends on several aspects, such as project type, environmental conditions, location, climate, available material, and economic considerations.

Among the range of additives, there is also a distinction into two categories, organic and chemical additives, representing the standard technologies for producing WMA mixtures [1–3] when referring to the use of additives. Since the construction of the first WMA pavements in 1997 in Germany with waxes as an additive, they have experienced considerable research leading to innovative solutions.

Waxes are organic additives incorporated in the asphalt mixture during the mixing process [4]. These additives are included in the asphalt binder and mixture to reduce the material viscosity. Some examples of different products used for WMA production in the category of organic additives can be found in the literature [5, 6]. More recently, Rodríguez-Alloza et al. [7] evaluated the combination of different additives and marginal materials. The authors investigated how waxes combined effect with other marginal materials on the rheological properties of asphalt binder. An increase in complex modulus was experienced for both waxes with a higher impact due to these additives. The addition of the marginal material also resulted in higher stiffness at low temperature. This effect that makes the binder more prone to thermal cracking is most likely due to the interaction with the WMA additives. Other studies were devoted to the fatigue behavior of WMA mixtures containing paraffin wax and synthetic wax [8, 9]. Together with enhanced fatigue resistance, both additives improved the workability, benefiting the material's performance overall. A recent research effort from Sedaghat et al. [10] has addressed the combined use of an organic WMA additive and nano-montmorillonite at the binder level. The rheological results indicated the beneficial effect of the nano-clay in improving the asphalt binder performance while limiting the potentially poorer behavior of the material at low temperature due to the presence of the organic additive.

Chemical additives represent an alternative WMA solution working as emulsifying agents promoting adhesion and coating of aggregate coating rather than as viscosity reducers that ultimately facilitate mixing and compaction at lower temperatures (5; 1; 2; 3). The combination of surfactants, polymers, or additives can improve coating properties, promote better mixture workability, and more effective compaction at lower temperatures. Nevertheless, the temperature reduction associated with chemical additives depends on the specific products (1). A study from Khani Sanij et al. [11] showed how a chemical additive improves several mechanical properties of asphalt binder and mixture, such as resilient modulus, creep, and moisture susceptibility. Other researchers [12, 13] supported the beneficial contribution of chemical additives in reducing compaction and production temperatures of asphalt mixture while enhancing the fatigue response. Furthermore, Sol-Sánchez et al. [14] stated that WMA from chemical additives could be produced in conventional asphalt plants without implementing significant changes in equipment, obtaining WMA with comparable performance to HMA while offering a cost-effective solution that can readily be incorporated into traditional plant procedure. In the same sense, other studies [15, 16] focused on assessing the performance of these WMA mixtures through trial sections and highlighted the viability of this technology, even when applied on rubberized mixtures for roads with high-volume traffic under severe climates.

A recent study by Habal and Singh [17] compared different WMA chemical and organic additives and their effects on the bonding of asphalt binder and aggregates. This investigation showed an aggregated dependence of the recycling agent's behavior with a relatively good correlation between bond strength and energy ratio. Different authors [18] also focused on evaluating chemical and organic WMA additives combined with other recycled materials. The experimentation highlighted the environmental benefits of the selected additives in terms of emission. However, poorer fatigue and thermal cracking performance were observed. A more extensive study on five different organic and chemical additives was proposed by Stienss and Szydlowsk [19] to evaluate WMA mixtures' low-temperature cracking susceptibility. The experimental results did not lead to a clear, conclusive outcome on the effect of the additives.

Concerning the foaming technology, this provides a diverse approach to the preparation of WMA material. In this method, the asphalt binder turns into foam when in contact with hot water [20]. The increase in volume due to the foaming process results in lower viscosity which facilitates aggregate coating and mixture workability. Two main types of foaming techniques exist: water-containing (indirect) and water-based (direct) methods (4; 1; 2; 3). Products from the first technique rely on fine water-containing powder added to the asphalt binder. After exceeding a specific temperature, this powder-binder blend foams due to the vaporization of water. On the other hand, dedicated equipment is used when operating with the direct (water-based) method. In this case, a more significant amount of water is directly added to the binder [2].

Wu et al. [21] evaluated several foaming technologies at the mixture level. They observed higher air voids and longitudinal cracking for WMA mixtures with transverse cracking and rutting response comparable to hot mix asphalt. Concerning the impact of the mixture constituents and testing conditions on the foamed WMA, Hou et al. [22] found that aggregate gradation and testing temperature influence field samples' behavior. This observation prompted the idea of developing dedicated job mix formula for WMA. Memon et al. [23] investigated the use of zeolite as an organic foaming agent. They identified an optimum zeolite content level of 3% to improve workability while limiting the energy demand during mixture production. An alternative and less expensive foaming additive named Alum (hydrated potassium aluminium sulphate) was studied in other research [24]. The WMA Alum-based material showed good resistance to thermal cracking and permanent deformation with satisfactory strength levels under dry and saturated conditions. Further research [25] focused on analyzing the fatigue cracking resistance of foaming WMA compared to the other technologies (organic and chemical additives). The results demonstrated that the WMA mixtures presented similar resistance to micro and macro-cracking development with respect to hot mix asphalt, with no significant differences between the various types of technologies evaluated.

An experimental and economic comparison between the conventional HMA with WMA mixtures made using the foaming process and organic additives was conducted by Gao et al. [26]. While the HMA presented a lower cost and overall superior mechanical performance, the foamed mixture exhibited better resistance to low temperature cracking and shear, with satisfactory moisture durability as compared to WMA mixture designed with organic additives. Chomicz-Kowalska et al. [27] investigated the simultaneous use of organic additive and foaming technology. Based on the experimentation, the authors observed a significant impact of the production temperature on the mixture performance leading to decreased compatibility and poorer mechanical behavior.

In a recent research effort, Sánchez et al. [28] addressed the use of foam WMA in the recycling process of RA. Dynamic Mechanical Analysis indicated that recycled binder, aging condition, and production temperature affect the material response. The following section provides more details on the use of WMA technology for pavement recycling.

3.2.2 *The Use of RAP in Combination with WMA*

The combination of WMA and RA is nowadays of great interest to the stakeholders of the road engineering field. This interest could be attributed to the environmental benefits derived from both the reduction in the temperature of the WMA and the ones from the reduction in consumption of virgin aggregates when substituted for RA.

The conjunction of both technologies leads to more resource-efficient and environmentally friendly asphalt pavements. Additionally, the drawbacks identified for different WMA technologies might be offset by the addition of RA, as RA helps increase mixture stiffness, especially when added in high amounts.

However, the literature points out concerns related to the mechanical performance of WMA + RA compared to traditional Hot Mix Asphalt (HMA) produced with fresh materials; the initially expected benefit due to the addition of high RA rates in WMA mixtures did not always improve the WMA, nor the HMA [29–31].

Concerning the decrease observed in the mechanical properties of the WMA + RA, the most significant ones are related to rutting susceptibility. This has been reported by Rathore et al. [32] when studying the performance of conventional HMA with virgin WMA and WMA containing 60% RA content. The virgin WMA mixture showed lower rutting resistance than HMA, and the addition of 60% RA content deteriorated the rutting performance of WMA. This latter effect was unexpected but also found by other authors [35]. However, other studies show how increased RA content improves rutting performance [30]. Nonetheless, it is also stated that an increase in the rejuvenator dosage, while permitting a higher amount of RA, can decrease the rutting performance due to an excessively softened aged asphalt binder. Thus, different types of WMA additives seem to affect rutting performance differently.

Regarding the moisture susceptibility of WMA + RA, some authors have identified that an increase in RA rates and rejuvenator dosage can reduce moisture resistance [33, 35]. In contrast, other studies found the opposite trend [34]. A recent research considering the addition of 60% RA in WMA did not find significant differences in the moisture resistance of these mixtures compared to virgin WMA and HMA [32]. These variabilities suggest that further research is needed when high RA and rejuvenator agents are used [30].

Another recent study [34] investigated the effect of long-term ageing on the properties of WMA + RA mixtures. The results showed that the WMA + RA retained much of its properties even after long-term ageing compared to HMA + RA mixes.

Most of the research dealing with WMA + RA has been performed at a laboratory scale, mainly focused on short-term properties. Regarding long-term performance, for those WMA with high rates of RA, some authors found a reduction in fatigue life and thermal properties [30].

3.3 Literature Review Summary

In summary, there is a consensus about the environmental benefits of combining WMA mixtures and RA. Amongst the recognized environmental advantages of WMA + RA mixtures is their contribution to the circular economy because of the reduction in consumption of virgin materials and the reutilization of the RA in the new mixture, and those derived from the reduction of the manufacturing temperature:

reduction in CO_2 emissions, energy consumption, and an improvement in health and safety.

Furthermore, the addition of RA to WMA was thought to provide a more sustainable material and improve some of the drawbacks of the WMA technologies. Specifically, the addition of RA provides a stiffer WMA, which might increase the WMA rutting resistance, and reduce the ageing of the binder in comparison to HMA. However, these improvements did not always occur, which might be due to the influence of the rejuvenators used, which is necessary for RA at high percentages, or if virgin binders that are too soft are selected.

During the last decade, most of the research related to WMA + RA focused on understanding its mechanical performance compared to HMA or WMA. Results showed significant variability and therefore, a general agreement in relation to the WMA + RA performance was not found. This can be due to the different WMA technologies applied (foaming, organics additives, and chemical additives) in the studies, the different types and content of additives, including ARA, different manufacturing temperatures and the variability of the RA used in the WMA + RA, that have been reported many times. The virgin binder selected for the WMA + RA and its content are also variable among various studies.

Regarding the long-term studies performed, although this is considered a key factor, few research results were reported, and it needs further investigation.

Finally, according to the literature, most of the research dealing with WMA + RA has been performed at a laboratory scale, and more information coming from field experience is needed.

3.4 Materials and Methods

3.4.1 General Principle

As part of this interlaboratory effort, six labs participated, named LAB1 to LAB6, from 6 countries (Canada, France, Germany, Italy, Spain and Switzerland). The experimental programme provided the comparison of physical and mechanical properties between a reference hot-mix asphalt (HMA) without RA and warm-mix asphalt (WMA), including 40% of RA by mix weight.

The objective of Task Group 2 was not focused on the properties of a particular bituminous material but on the effect of the combination of WMA + RA on the properties of the bituminous material in general. Therefore, it was initially decided not to choose a specific common material for the inter-lab testing as usual for a round-robin test. By contrast, the use of a WMA process and the recycling of RA are required. This implies that the results of the laboratories on different materials are not comparable in absolute values. For this reason, parallel measurements on a hot mix asphalt with the same design parameters taken as local reference are also required. In this way, the relative effect of the combination of WMA + RA on the mix properties can be calculated and compared between the different laboratories. However, the causes must also be similar for the effects to be comparable. Consequently, it was also decided to use the same RA material, the same virgin binders, and the same recycling rate fixed at 40%.

3.4.2 Characterization of the Common RA and Virgin Binders

In the round-robin test, the RA and the bituminous binders were the same for all the laboratories. In particular, the RA was collected from a French road and separated in batches using an industrial quartering machine to ensure a good homogeneity. Finally, it was shipped to all the participating labs in 25 kg bags. The selected binders were 50/70 and 160/220 penetration binders, to be used respectively for HMA and WMA mixtures. Finally, each laboratory used different virgin aggregates and warm-mix technology (additive and/or foam).

In a preliminary step of the research, the common materials were independently characterized by the participating laboratories. Figures 3.1, 3.2, 3.3 show the RA binder content, the RA aggregate gradation (white curve), and the properties of the virgin and RA binders, respectively.

The results obtained show some significant differences between the laboratories. In particular, all the laboratories determine a RA binder content of approximately 4.5% by RA weight, except for Lab#3 (Fig. 3.1). Indeed, the result from Lab#3 is insignificant, as it is more than six times higher than the average estimated reproducibility values mentioned in the standard EN 12,697–1 (R = 0.39%). Similarly, the RA aggregate gradation showed relevant differences among the different laboratories, considering the R value provided in the standard EN 12,697–2 (R = 1.7%).

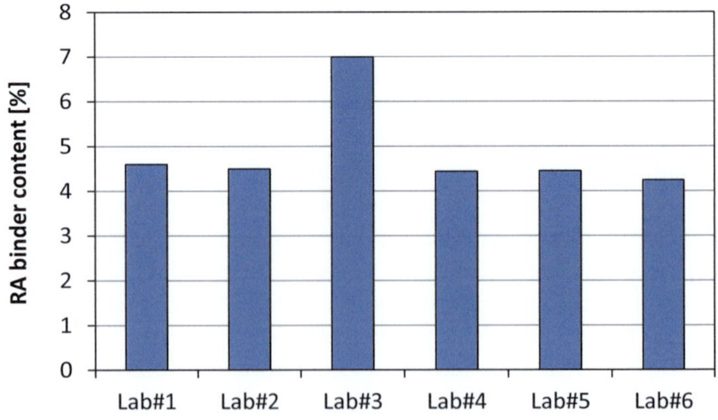

Fig. 3.1 Binder content in the RA

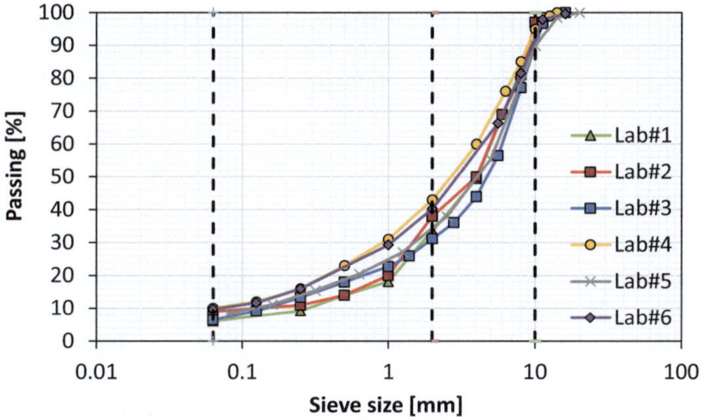

Fig. 3.2 Gradation of the RA aggregate particles (white curves)

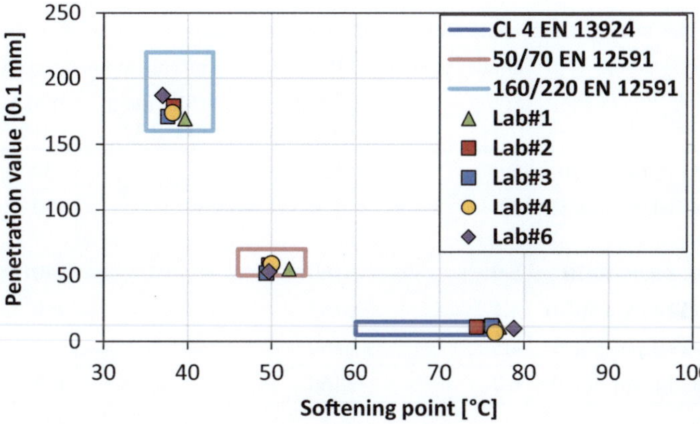

Fig. 3.3 Properties of the virgin and RA binders

In particular, the ranges of the passing at 10 mm, 2 mm and 0.063 mm sieves were respectively 7.6%, 11.9% and 3.9% (Fig. 3.2). These differences can be ascribed to heterogeneities between the RA batches or the procedures adopted by the laboratories. Though an industrial quartering machine was used for the RA sampling, it is impossible to specifically attribute the differences to the first or the second cause. In the inter-laboratory testing program framework, the RA characterization step was initially intended to be a routine preparation for the mix design step. The results show that this characterization is not as straightforward as it may appear. Nevertheless, it is a key element for the mix design in the recycling process as it determines the relevance of all the following stages.

A good data superposition was obtained in terms of penetration value and softening point of the virgin and RA binders (Fig. 3.3). In particular, the data from the virgin

50/70 and 160/220 binders perfectly comply with the ranges provided by EN 12,591. The RA binder provided penetration values of about 7–12 dmm, matching the class 4 defined by EN 13,924, but showed softening points near or higher than the maximum.

3.4.3 Mixes Design

Table 3.1 shows the mix types selected by the six participating laboratories, according to EN 13,108–1. The corresponding grading curves are plotted in Fig. 3.4.

Overall, the materials encompass a range of mixture types from base to surface courses and the two types of warm processes: WMA with foam (two laboratories) and WMA with additive (5 laboratories). For four laboratories, the warm technology allowed decreasing the mixing temperature of 20 °C, while in the case of Lab#4 and Lab#6, the reduction was 25 °C. It has to be noted that the mixtures manufactured

Table 3.1 Types of mixes, processes and mixing temperatures chosen by the laboratories

Lab	Designation according to EN 13,108–1	Binder content (%)	Processes and mixing temperatures (°C)		
			HMA (0% RA)	WMA foam (40% RA)	WMA additive (40% RA)
Lab#1	AC 16 bin 50/70	5.0	150	–	130
Lab#2	AC 10 surf/bin 50/70	5.4	150	130	130
Lab#3	AC 8 surf 50/70	5.9	160	–	140
Lab#4	AC 14 base 50/70	4.1	155	–	130
Lab#5	AC 10 surf 50/70	4.5	160	140	–
Lab#6	AC 11 surf 50/70	6.2	160	–	135

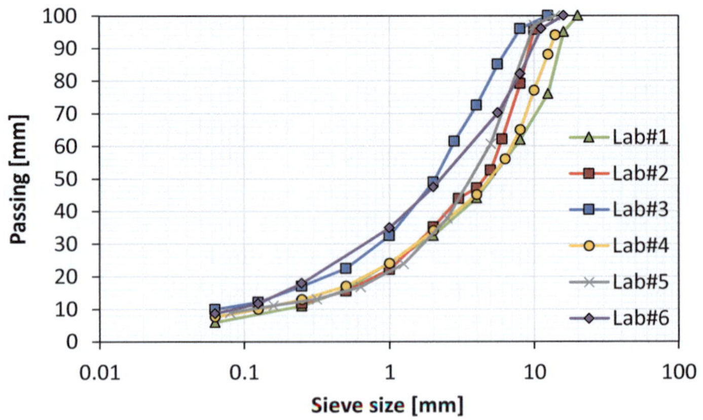

Fig. 3.4 Grading curves of the mixes chosen by the laboratories

by the six laboratories show significant differences in terms of gradation, maximum aggregate size and total binder content. In particular, the latter resulted in different ratios between aged binder from RA (which is constant, as the RA material was the same for all the laboratories and the RA content in the mix was fixed at 40%) and virgin binder. These differences may have determined heterogeneous results among the different laboratories, even if the interlaboratory test approach provided a comparison between HMA and WMA with RA under the same conditions.

3.5 Key Findings

3.5.1 Tests Performed and Results

The properties studied by the participating laboratories are summarized in Table 3.2, while the results obtained for each property are shown in Tables 3.3, 3.4, 3.5, 3.6, 3.7. According to the type of test, the relative effect is calculated with the ratio WMA/HMA if high values reflect better properties (water sensitivity, modulus, and fatigue) and with the ratio HMA/WMA if low values reflect better properties (compactability and rutting). So, relative values higher than 1 indicate a better performance for the WMA; vice versa, relative values lower than 1 indicate better performance for the HMA.

All the laboratories that analysed the mix compactability compared the air voids content of gyratory compacted specimens at a specific number of gyrations. The results in terms of air voids content ranged between 1.5% and 9.2%, proving the wide data variability among the different laboratories. However, the relative effect HMA/WMA varied between 0.94 and 1.13. This indicates that the presence of RA and the temperature reduction did not significantly affect the mix compactability, denoting good effectiveness of the warm processes (additive and foam).

The laboratories assessed the water sensitivity of the mixtures in terms of the indirect tensile strength ratio (ITSR) or compressive strength ratio (i/C). All the laboratories obtained water sensitivity values between 85 and 90% for the HMA mix. Higher variability was observed for the WMA mixes, including RA, which

Table 3.2 Properties investigated in the laboratories

Property	HMA	WMA foam (40% RA)	WMA additive (40% RA)
Compactability	Lab#1, 2, 5, 6	Lab#2, 5	Lab#1, 2, 6
Water sensitivity	Lab#1, 2, 4, 5, 6	Lab#2, 5	Lab#1, 2, 4, 6
Rutting	Lab#1, 2, 3, 5	Lab#2, 5	Lab#1, 2, 3
Modulus	Lab#1, 2, 5	Lab#5	Lab#1, 2
Fatigue	Lab#1, 2, 6	Lab#2	Lab#1, 2, 6

Table 3.3 Compactability test results

Lab	Test standard	Parameter	HMA (%)	WMA foam (40% RA)		WMA additive (40% RA)	
				Absolute	Relative effect (HMA/WMA)	Absolute	Relative effect (HMA/WMA)
Lab#1	EN12697-31	Voids @ 100 gyrations	4.2	–	–	4.0%	1.05
Lab#2	EN12697-31	Voids @ 60 gyrations	9.2	8.4%	1.10	8.9%	1.03
Lab#5	EN12697-31	Voids @ 80 gyrations	6.6	7.0%	0.94	–	–
Lab#6	EN12697-31	Voids @ 100 gyrations	1.7	–	–	1.5%	1.13

Table 3.4 Water sensitivity test results (EN 12,697–12 A)

Lab	Test standard	Parameter	HMA (%)	WMA foam (40% RA)		WMA additive (40% RA)	
				Absolute	Relative effect (WMA/HMA)	Absolute	Relative effect (WMA/HMA)
Lab#1	EN12697-12 A	ITSR	88.4			99.6%	1.13
Lab#2	EN12697-12 A	ITSR	86.6	86.3%	1.00	91.9%	1.06
Lab#2	EN12697-12 B	i/C	86.7	95.9%	1.11	98.5%	1.14
Lab#4	EN12697-12 A	ITSR	87.0	–	–	88.6%	1.02
Lab#4	EN12697-12 B	i/C	89.3	–	–	88.1%	0.99
Lab#5	EN12697-12 A	ITSR	85.7	79.4%	0.93	–	–
Lab#6	EN12697-12 A	ITSR	88.2	–	–	92.6%	1.05

shows ITSR or i/C values ranging between 79.4% and 99.6%. The relative effect of WMA/HMA varied from 0.93 to 1.14, indicating that the WMA mixes, either produced using additive or foam, did not show any particular weakness to water damage.

Table 3.5 Rutting test results

Lab	Test standard	Parameter	HMA (%)	WMA foam (40% RA)		WMA additive (40% RA)	
				Absolute	Relative effect (HMA/WMA)	Absolute	Relative effect (HMA/WMA)
Lab#1	EN12697-25 B	$\varepsilon_{1000,calc}$	1.64	–	–	1.27%	1.29
Lab#2	EN12697-22	d @ 30,000 cycles	4.6	4.6%	1.00	4.8%	0.96
Lab#3	EN12697-22	d @ 10,000 cycles	5.9	–	–	1.8%	3.28
Lab#5	EN12697-22	d @ 10,000 cycles	6.0	7.7%	0.78		
Lab#5	EN12697-22	d @ 30,000 cycles	7.4	10.7%	0.69	–	–

Table 3.6 Modulus test results

Lab	Test standard	Parameter	HMA	WMA foam (40% RA)		WMA additive (40% RA)	
				Absolute	Relative effect (WMA/HMA)	Absolute	Relative effect (WMA/HMA)
Lab#1	AASHTO T341-11	$\|E^*\|$ @ 50°C, 0.1 Hz	144 MPa	–	–	127 MPa	0.88
Lab#1	AASHTO T341-11	$\|E^*\|$ @ 20°C, 1 Hz	5380 MPa	–	–	4981 MPa	0.93
Lab#1	AASHTO T341-11	$\|E^*\|$ @ 5°C, 20 Hz	28,175 MPa	–	–	23,077 MPa	0.82
Lab#2	EN12697-26	$\|E^*\|$ @ 15°C, 10 Hz	9103 MPa	–	–	7726 MPa	0.85
Lab#5	AASHTO T341-11	$\|E^*\|$ @ 35°C, 0.1 Hz	295 MPa	101 MPa	0.34	–	–
Lab#5	AASHTO T341-11	$\|E^*\|$ @ 5°C, 1 Hz	16,426 MPa	9357 MPa	0.57	–	–
Lab#5	AASHTO T341-11	$\|E^*\|$@-35°C, 10 Hz	36,483 MPa	33,452 MPa	0.92	–	–

Table 3.7 Fatigue test results

Lab	Test standard	Parameter	HMA	WMA foam (40% RA)		WMA additive (40% RA)	
				Absolute	Relative effect (WMA/HMA)	Absolute	Relative effect (WMA/HMA)
Lab#1	AASHTO TP107-18	N_f @ 400 με	52,000	–	–	175,000	3.37
Lab#1	AASHTO TP107-18	N_f @ 600 με	13,500			52,500	3.89
Lab#6	EN12697-24	$N_{f,50}$	12,397	–	–	12,684	1.02

The rutting tests showed very heterogeneous results. In Lab#5, the WMA, including RA, had a rut depth compared to HMA, with relative values minor than 0.8. On the contrary, in Lab#1 and Lab#3, resistance to permanent deformation was higher for the WMA with RA. Finally, Lab#2 had similar rut depth values for the HMA, WMA-foam and WMA-additive mixtures. The different mix types, gradation, and binder content probably hinder the laboratories from obtaining a univocal result.

The results from the modulus tests showed, in general, a stiffer behaviour for HMA compared to WMA. For Lab #1 and Lab #2, the relative effect of WMA-additive/HMA was in the interval 0.82–0.93, indicating a modulus reduction of approximately 10–15% at all the explored temperatures and frequencies. For Lab#5, the relative difference between HMA and WMA-foam increased when reducing the frequency and increasing temperature. This result is probably related to the softer virgin binder (160/220 penetration grade binder) used in the WMA mixtures, which over-compensated the hardness of the RA binder, and/or to the less detrimental short-term ageing of the binder for the warm processes due to the temperature reduction.

For the fatigue test, the data set is very limited (only two laboratories). Moreover, Lab#1 observed a higher resistance to fatigue for the WMA, while for Lab#6, the performance of HMA and WMA was comparable (relative effect WMA/HMA of 1.02). Due to the limited availability of data and the heterogeneities in the results, an assessment of the WMA's fatigue behavior needs further research.

3.6 Conclusions

The Task Group 2 of the RILEM Technical Committee 264-RAP on non-cold recycling identified the current trend in the use of WMA + RA and the need to bring some insights about various durability issues associated with the WMA + RA combination. For this purpose, an international inter-laboratory testing program was proposed in which six different laboratories were involved. All the participants carried out a comparative study between an HMA and a WMA + 40% RA (considering two processes: foaming and chemical additives). The properties studied for the mechanical properties analysis were compactability, moisture and rutting resistance, stiffness modulus and fatigue.

The results obtained, which were limited, did not identify any specific issues for the use of WMA + 40%RA compared to the HMA.

Nonetheless, some considerations related to the inter-laboratory testing program, such as the limited data set, are more significant for fatigue testing.

One question to consider is how representative the laboratory mixing process could be compared to the different plant processes. In continuous plants, the RA can be heated through direct (in a doble drum) or indirect ways (by contact with overheated virgin aggregates). In contrast, in batch plants, the RA can be indirectly heated and directly heated but also directly heated and mixed with rejuvenators.

Based on the results obtained in this research, some recommendations for practice are suggested:

- A reliable characterization of the RA is needed; it is a prerequisite for successful recycling.
- Considering the addition of high amounts of RA, compensation for the RA aged binder is needed. In this sense, the use of soft virgin binders to compensate for the ageing of the RA binder has proved to be efficient.

Finally, several research needs have been identified through the efforts of Task Group 2 of the RILEM TC 264-RAP. Long-term performance tracking is necessary to better understand the durability of warm mix asphalt with recycled asphalt (WMA + RA). More information from the field trials is also required as most research projects have focused on laboratory evaluation. Another issue is related to recyclability, considering that multi-recycling is becoming prominent, it is also necessary to conduct studies on the recyclability and the compatibility of the different products.

Acknowledgements The RILEM Technical Committee on Asphalt Pavement Recycling (TC 264-RAP) and Task Group 2 Members are gratefully acknowledged. The authors would also like to thank Nynas AB and Université Gustave Eiffel (formerly IFSTTAR) for supporting the interlaboratory activity by providing asphalt binders and RA mixtures.

References

1. Cheraghian, G., Cannone, F.A., You, Z., Chen, S., Kim, Y.S., Westerhoff, J., Moon, K.H., Wistuba, M.: Warm mix asphalt technology: An up to date review. J. Clean. Prod. **268**, 122128 (2020)
2. Behnood, A.: A review of the warm mix asphalt (WMA) technologies: Effects on thermo-mechanical and rheological properties. J. Clean. Prod. **259**, 122817 (2020)
3. Sukhija, M., Saboo, N.: A comprehensive review of warm mix asphalt mixtures-laboratory to field (2021)

4. Zaumanis, M.: Warm mix asphalt investigation. Master of science thesis. Kgs. Lyngby: technical university of denmark in cooperation with the danish road institute, department of civil engineering (2010)
5. Rubio, M.C., Martínez, G., Baena, L., Moreno, F.: Warm mix asphalt: an overview. J. Clean. Prod. **24**, 76–84 (2012)
6. Ji, J., Yao, H., Yuan, Z., Suo, Z., Xu, Y., Li, P., You, Z.: Moisture susceptibility of warm mix asphalt (WMA) with an organic wax additive based on X-ray computed tomography (CT) technology. Adv. Civ. Eng. (2019)
7. Rodríguez-Alloza, A.M., Gallego, J.: Mechanical performance of asphalt rubber mixtures with warm mix asphalt additives. Mater. Struct. **50**(2), 147 (2017)
8. Leng, Z., Yu, H., Zhang, Z., Tan, Z.: Optimizing the mixing procedure of warm asphalt rubber with wax-based additives through mechanism investigation and performance characterization. Constr. Build. Mater. **144**, 291–299 (2017)
9. Yu, H., Leng, Z., Dong, Z., Tan, Z., Guo, F., Yan, J.: Workability and mechanical property characterization of asphalt rubber mixtures modified with various warm mix asphalt additives. Constr. Build. Mater. **175**, 392–401 (2018)
10. Sedaghat, B., Taherrian, R., Ali, H.S., Mojtaba, M.S.: Rheological properties of bitumen containing nanoclay and organic warm-mix asphalt additives. Constr. Build. Mater. **243**, 118092 (2020)
11. Khani, H., Meybodi, P.A., Hormozaky, M.A., Hosseini, S.H., Olazar, M.: Evaluation of performance and moisture sensitivity of glass-containing warm mix asphalt modified with zycothermTM as an anti-stripping additive. Constr. Build. Mater. **197**, 185–194 (2019)
12. Pereira, R., Almeida-Costa, A., Duarte, C., Benta, A.: Warm mix asphalt: Chemical additives' effects on bitumen properties and limestone aggregates mixture compactibility. Int. J. Pavement Res. Technol. **11**(3), 285–299 (2018)
13. Kumar, A., Choudhary, R., Kandhal, P.S., Julaganti, A., Behera, O.P., Singh, A., Kumar, R.: Fatigue characterisation of modified asphalt binders containing warm mix asphalt additives. Road Mater. Pavement Des, 1–23 (2018)
14. Sol-Sánchez, M., Moreno-Navarro, F., Rubio-Gámez, M.C.: Study of surfactant additives for the manufacturing of Warm Mix Asphalt: from laboratory design to asphalt plan manufacture. Appl. Sci. **7**(7), 745 (2017)
15. Sol-Sánchez, M., Moreno-Navarro, F., García-Travé, G., Rubio-Gámez, M.C.: Analysing industrial manufacturing in-plant and in-service performance of asphalt mixtures cleaner technologies. J. Clean. Prod. **121**, 56–63 (2016)
16. Moreno-Navarro, F., Sierra, F.J., Sol-Sánchez, M., Rubio-Gámez, M.C., Castillo-Garzón, M., Estévez, E.: High-performance sustainable asphalt mixtures for high-volume track roads in severe climates. Sustainability **12**(21), 8765 (2020)
17. Habal, A., Singh, D.: Effects of warm mix asphalt additives on bonding potential and failure pattern of asphalt-aggregate systems using strength and energy parameters. Int. J. Pavement Eng. **22**(4), 467–479 (2021)
18. Pouranian, M., Ali, N.M., Tabesh, M.T., Nazeri, B., Shishehbor, M.: Rheological and environmental characteristics of crumb rubber asphalt binders containing non-foaming warm mix asphalt additives. Constr. Build. Mater. **238**, 117707 (2020)
19. Stienss, M., Szydlowsk, C.: Influence of selected warm mix asphalt additives on cracking susceptibility of asphalt mixtures. Mater. **13**, 202 (2020)
20. Hasan, M.R.M., You, Z., Yin, H., You, L., Zhang, R.: Characterizations of foamed asphalt binders prepared using combinations of physical and chemical foaming agents. Constr. Build. Mater. **204**, 94–104 (2019)
21. Wu, S., Wen, H., Zhang, W., Shen, S., Mohammad, L.N., Faheem, A., Muhunthan, B.: Field performance of top-down fatigue cracking for warm mix asphalt pavements. Int. J. Pavement Eng. **20**(1), 33–43 (2019)
22. Hou, Y., Cai, Y., Zang, Z., Qian, Z., Zhao, B.: Dynamic characteristics of warm mix foamed asphalt mixture in seasonal frozen area. Adv. Mater. Sci. Eng. (2019)

23. Memon, A., Hazwani, N., Hartadi, S.M., Bala, N.: Rheological properties of bitumen containing organic foaming agent for warm mix asphalt. Earth Environ. Sci. **498**, 012028 (2020)
24. Mosa, A.M., Salem, L.A., Waryosh, W.A.: New admixture for foamed warm mix asphalt: a comparative study. Iran J Sci Technol, Trans. Civ. Eng. **44**, 649–660 (2020)
25. Sol-Sánchez, M., Fiume, A., Moreno-Navarro, F., Rubio-Gámez, M.C.: Analysis of fatigue cracking of warm mix asphalt. Influence of the manufacturing technology. Int. J. Fatigue **110**, 197–203 (2018)
26. Gao, K., Chen, F., Kuai, H., Yang, Z., Si, Y., Huo, Z.: Laboratory experimental research on foamed warm mix asphalt mixture road property. Mater. Sci. Eng. **774**, 012072 (2020)
27. Chomicz-Kowalska, A., Maciejewski, K., Iwanski, M.M.: Study of the simultaneous utilization of mechanical water foaming and zeolites and their effects on the properties of warm mix asphalt concrete. Mater. **12**(2), 357–379 (2020)
28. Sánchez, D.B., Airey, G., Caro, S., Grenfell, J.: Effect of foaming technique and mixing temperature on the rheological characteristics of fine RAP-foamed bitumen mixtures. Road. Mater. Pavement Des. **21**(8), 2143–2159 (2020)
29. Farooq, M., Mir, M.: Use of reclaimed asphalt pavement (RAP) in warm mix asphalt (WMA) pavements: a review. Innov. Infrastruct. Solut, 10 (2017)
30. Guo, M., Liu, H., Jiao, Y., Mo, L., Tan, Y., Wang, D., Liang, M.: Effect of WMA-RAP technology on pavement performance of asphalt mixture: A state-of-the-art review. Constr. Build. Mater. **266**, 121704 (2020)
31. Yousefi, A., Behnood, A., Nouruzi, A., Haghshenas, H.: Performance evaluation of asphalt mixtures containing warm mix asphalt (WMA) additives and reclaimed asphalt pavement (RAP). Constr. Build. Mater. **268**, 121200 (2021)
32. Rathore, M., Haritonovs, V., Zaumanis, M.: Performance evaluation of warm asphalt mixtures containing chemical additives and effect of incorporating high reclaimed asphalt content. Mater. **14**, 3793 (2021)
33. Goli, H., Latifi, M.: Evaluation of the effect of moisture on behavior of warm mix asphalt (WMA) mixtures containing recycled asphalt pavement (RAP). Constr. Build. Mater. **247**, 118526 (2020)
34. Monu, K., Ransinchung, G.D.R.N., Surender, S.: Effect of long-term ageing on properties of RAP inclusive WMA mixes. Constr. Build. Mater. **206**, 483–493 (2019)
35. Zhao, S., Huang, B., Shu, X., Woods, M.: Comparative evaluation of warm mix asphalt containing high percentages of reclaimed asphalt pavement. Constr. Build. Mater. **44**, 92–100 (2013)

Chapter 4
Asphalt Binder for Recycled Asphalt Mixtures

Augusto Cannone Falchetto, Martin Hugener, Di Wang, Laurent Porot, Patricia Kara De Maeijer, Marko Orešković, Margarida Sa-da-Costa, Hassan Tabatabaee, Edoardo Bocci, Atsushi Kawakami, Bernhard Hofko, Andrea Grilli, Emiliano Pasquini, Marco Pasetto, Huachun Zhai, Hilde Soenen, Wim Van den Bergh, Fabrizio Cardone, Alan Carter, Kamilla Vasconcelos, Xavier Carbonneau, Eshan Dave, and Gabriele Tebaldi

A. C. Falchetto (✉)
Department—University, ZIP, USA
e-mail: augusto.cannone@unipd.it

M. Hugener
Empa—Materials Science and Technology, Überlandstrasse, Dübendorf, Switzerland
e-mail: martin.hugener@empa.ch

D. Wang
Braunschweig Pavement Engineering Centre, Technische Universität Braunschweig, Braunschweig, Germany
e-mail: di.wang@tu-bs.de

L. Porot
Kraton Chemical B.V, Transistorstraat, CE Almere, Netherlands
e-mail: laurent.porot@kraton.com

P. K. De Maeijer · W. Van den Bergh
EMIB, Faculty of Applied Engineering, University of Antwerp, Antwerp, Belgium
e-mail: patricija.karademaeijer@uantwerpen.be

W. Van den Bergh
e-mail: wim.vandenbergh@uantwerpen.be

M. Orešković
University of Belgrade, Faculty of Civil Engineering, Bulevar Kralja Aleksandra, Belgrade, Serbia
e-mail: moreskovic@grf.bg.ac.rs

M. Sa-da-Costa
National Laboratory for Civil Engineering, Lisboa, Portugal
e-mail: mcosta@lnec.pt

G. Tebaldi and E. Dave (eds.), *Recycling Asphalt Pavement*, RILEM State-of-the-Art Reports 44, https://doi.org/10.1007/978-3-031-94231-0_4

Keywords Reclaimed Asphalt Binder · Rejuvenators · Binder Blending · Ageing Behaviour · Asphalt Recycling · Interlaboratory Testing

---

H. Tabatabaee
Cargill, Wayzata, Minneapolis, MN, USA
e-mail: hassan_tabatabaee@cargill.com

E. Bocci
eCampus University, Novedrate (CO), Italy
e-mail: edoardo.bocci@uniecampus.it

A. Kawakami
Department of Biological Information, Tokyo Institute of Technology, Midori-Ku, NagatsutaYokohama, Japan
e-mail: atkawaka@bio.titech.ac.jp

B. Hofko
Institute of Transportation, TU Wien, Gusshausstrasse, Vienna, Austria
e-mail: bernhard.hofko@tuwien.ac.at

A. Grilli
Department of Economics, Science and Law, University of the Republic of San Marino, San Marino, Republic of San Marino
e-mail: andrea.grilli@unirsm.sm

E. Pasquini · M. Pasetto
Department of Civil, Environmental and Architectural Engineering (ICEA), University of Padua, Padua, Italy
e-mail: emiliano.pasquini@unipd.it

M. Pasetto
e-mail: marco.pasetto@unipd.it

H. Zhai
Idaho Asphalt Supply Inc, Nampa, USA
e-mail: hzhai@idahoasphalt.com

H. Soenen
NYNAS NV, Bitumen Research, Antwerp, Belgium
e-mail: hilde.soenen@nynas.com

F. Cardone
Department of Civil and Building Engineering and Architecture, Università Politecnica Delle Marche, Via Brecce Bianche, Ancona, Italy
e-mail: f.cardone@staff.univpm.it

A. Carter
Department of Construction Engineering, École de Technologie Supérieure (LCMB-ETS), Montreal, QC, Canada
e-mail: alan.carter@etsmtl.ca

K. Vasconcelos
Laboratory of Pavement Technology, University of São Paulo, Cidade Universitária, São Paulo, SP, Brazil
e-mail: kamilla.vasconcelos@usp.br

4.1 Introduction

Asphalt binder is an essential component of asphalt mixture for pavement applications [1]. This is also true in the case of mixtures that contain reclaimed asphalt (RA) since the different physical and chemical characteristics of the binder from the RA particles and the properties of the virgin binder (Vb) strongly influence the response of the mixture [2]. Over the years, different solutions have been investigated to improve the recycling process and increase the amount of RA in the asphalt mix design [3]. For example, this includes (i) the use of a softer binder to re-establish the appropriate consistency of the binder phase in the mixture using RA [4] (this approach presents a series of limitations); and (ii) the incorporation of additives or agents capable of reinstating, fully or partially, the properties of the binder contained in the RA when blended with Vb.

In this sense, it is worth introducing and justifying two terms that are adopted in this chapter. The acronym RA, commonly accepted in Europe, stands for Reclaimed Asphalt [5], and it is selected in place of RAP, which is conventionally adopted in the USA to identify Reclaimed Asphalt Pavement material (also known as *recycled asphalt pavement, RAP*) [6]. It should be noted that there is an ongoing debate considering a change of paradigm from the concept of recycling to re-use, broadening the application of the material toward a more solid circular economy and waste management hierarchy while highlighting the potential added value of the reclaimed material [7].

Considering the rejuvenators, new terminology is proposed to identify this class of materials: Asphalt Recycling Agent (ARA). The latter focuses the attention on the materials used for asphalt application purposes and appears to be more specific for asphalt materials. A more general term was also considered: Asphalt Mixture and Binder Recycling/Re-use Agent (AMBRA). However, ARA was finally adopted by RILEM TC 264-RAP Task Group 3, as no asphalt mixture was investigated in this research.

Recycling and reusing asphalt pavement is vital not only because of the potential environmental, economic, and societal benefits (also addressed as the 4Ps "benefits: Planet, People, Profitability and Performance") associated with this process but also because of the significant scale of these benefits since milled asphalt pavement together with crushed concrete and bricks is one of the main materials obtained

X. Carbonneau
Campus Scientifique Et Technique (CST) COLAS, Magny-Les-Hameaux, France
e-mail: carbonneau@campus.colas.fr

E. Dave
Department of Civil and Environmental Engineering, University of New Hampshire, Durham, NH, USA
e-mail: eshan.dave@unh.edu

G. Tebaldi
Department of Engineering and Architecture, University of Parma, Parma, Italy
e-mail: gabriele.tebaldi@unipr.it

from Construction and Demolition Waste (CDW) [8]. Gaining a more in-depth understanding of the use of ARA can have a favourable impact on the recycling process of RA, further encouraging their use by engineers, practitioners, construction companies, road authorities, and departments of transportation. In this context, a sizeable interlaboratory activity at the international level in the RILEM circuit can provide strong and solid support toward a comprehensive evaluation of the use of ARA starting from the bitumen level. In this sense, the RILEM research activities in the field of characterization of bituminous materials and recycled asphalt pavement started at the end of the last century (TC 017-BM, TC 101-BAT, TC 152-PBM, TC 182-PEB). The actions of the TC 264-RAP Task Group 3 are the natural continuation and extension of the active research impulse originating from the seminal works performed in TC 206-ATB [9], TC 237-SIB [10], and TC 252-CMB [11–13]. From this perspective, the knowledge of previous TCs in terms of testing, characterization, rheological and chemical analysis, and recycling approach and ageing phenomena were combined with a clear focus on the asphalt binder phase and its interaction with ARA.

4.2 Literature Review

In most countries, the development of the reuse and recycling of asphalt pavements started after the energy crisis in 1973 [9, 14, 15]. Recycling was limited to a maximum of 15 to 20% RA. It is known that using high contents of RA in asphalt mixtures increases the benefits of recycling but at the same time, also limits it due to the ageing of asphalt binder [6, 16, 17]. On the one hand, the oxidative ageing mechanism of asphalt binder is complex and can result in adverse effects, especially in the fatigue and thermal properties showing brittle behaviour [18]. On the other hand, the material becomes stiffer as it ages [19], which can positively impact the rutting resistance of the pavement. This represents a considerable constraint for pavement engineers and road authorities as more and more RA will have to be re-used and recycled in the upcoming years, possibly affecting pavement performance beyond acceptable service levels. At the same time, as the RA is from 10 to 20 years old roads, the RA for tomorrow will be different under the impetus of more use of harder bitumen grades, Polymer Modified Bitumen (PMB) or second or third recycling cycles. This trend has already been observed in countries like Japan or Germany [20, 21]. Hence, a technology that aims to use higher RA percentages or RA with a harder aged binder must be able to restore the lost properties of the aged binder without affecting the benefits gained with ageing [22, 23].

ARAs are one such technology. ARAs are additives used to improve the characteristics of asphalt mixtures, including a high content of RA or hard RA binder in a way to possibly restore the physical properties of aged binder and ensure long-lasting asphalt concrete [24, 25]. Adding an ARA to a RA mixture can be a potential path to increase the pavement's service life while also improving sustainability. Several studies have shown that some ARAs can substantially restore the properties of aged asphalt binder with a limited negative impact on the binder's rheological properties and temperature susceptibility [26–29]. In recycled mixtures, both rejuvenation and ageing occur from the outside to the inside of the RA particles: the outer layer, which is most affected by the ageing process, receives the rejuvenator benefits earlier than the inner layer, which experiences less ageing and hence a less remarkable rejuvenating effect. Therefore, to some extent, these two processes balance each other and deliver a relatively homogeneous asphalt binder. It can be considered an advantage of recycled mixtures over virgin mixtures which are less homogeneous due to inconsistent ageing across the binder film [16]. Considering short-term ageing (STA) and long-term ageing (LTA), the service life of the binder can be simulated by using the rolling thin film oven test (RTFOT) and pressure ageing vessel (PAV) testing procedures under ASTM [30, 31] and CEN [32, 33] standards.

Different types of ARAs may have diverse effects on the aged binder and usage [34–39]. ARAs involve interactions between the different chemical fractions in bituminous binders while improving damage resistance and maintaining that impact with ageing. Proper rejuvenation improves the processability of RA asphalt mixtures, cracking resistance and durability without compromising on rutting resistance [40]. Some ARAs, in high-amount RA mixtures might avoid increasing the production and compaction temperature of the mixtures. With hot recycling (simulated by blending LTA with Vb), the viscous response of the binder is enhanced when ARAs are added [41–43]. Careful selection of ARAs is required to provide the pavement with the necessary short and long-term properties [44]. In STA criteria, ARAs must diffuse rapidly into the RA binder (RAb) and mobilize the aged asphalt. In LTA criteria, the chemical structure should be altered in addition to binder rheology to minimize fatigue and low-temperature cracking [45]. The optimum ARA dosage depends not only on ARA type and the total amount of RA but also on the ageing level the RAb has undergone [46]. Considering that it is usually not possible to achieve the restoration of all rheological properties when rejuvenating an aged binder, it is necessary to perform a multi-criteria evaluation to determine the optimum doses of ARAs [47].Several researchers have focused on the performance of the rejuvenated binder or bituminous blends with re-ageing, with different outcomes [26, 48–55]. Ongel and Hugener [53] studied the ageing behaviour of a 40/50 Vb (used to simulate the characteristics of a RAb) with three types of ARAs (dosage varied from 5.0% to 6.0% by aged binder weight) and compared it with that of the 70/100 Vb. The research showed that the ageing of the Vb is slower than that of the rejuvenated binder; therefore, the rejuvenated binder may not perform as well as the virgin binder in the LTA. Grilli et al. [26] tested a blend of laboratory-aged binder (RTFOT + PAV) and ARAs (with two dosages of 3% and 6% by aged binder weight) and analysed the effect of re-ageing (RTFOT + PAV). The ageing effect on rejuvenated binders was noted to be slightly more detrimental than that on the Vb since it entailed a higher variation of rheological parameters. Yin et al. [54] focused on characterizing the ARAs effectiveness on bituminous blends and mixtures with high RA and RAS contents, carrying out laboratory tests at different ageing levels. The results indicated that the ARAs evaluated in this study (two types with dosages of 4.5% and 5.5% by aged binder weight, respectively) were effective in partially restoring the properties of recycled materials, but their rejuvenating effectiveness diminished with ageing. Nevertheless, the recycled blends and mixtures with ARAs achieved comparable or even better rheological properties and cracking resistance determined by the Illinois flexibility index test (I-FIT) than those with an allowable number of recycled materials per agency specifications but without ARA. Mazzoni et al. [52] investigated the evolution of the rheological properties of the binder with ageing (first service life), hot recycling (blending of virgin and aged binder, without adding ARA or adding with three different ARA), and re-ageing (service life after RA hot recycling). The results showed that the hot recycled bituminous blends, including a properly selected and dosed ARA (6% by aged binder weight), can undergo less severe oxidation than Vb and can be less stiff at the end of the service life. Nsengiyumva et al. [37] have investigated the effects of the type, dosage (7%—16.1%) and treatment methods of ARA on chemical–mechanical characteristics of high-RA mixtures. The results of saturates-aromatics-asphaltenes-resin (SARA) analysis showed that the laboratory ageing increased asphaltenes whereas the aromatics decreased; the addition of rejuvenators to the aged binder diluted the asphaltene content, and changes in aromatics and resins were ARA dependent. Bocci et al. [49] studied the evolution of the rheological behaviour of the binder extracted from RA with rejuvenation (blending of Vb, RAb, and ARA) and re-ageing. The results proved that, after re-ageing (RTFOT + PAV), the rejuvenated RAb showed a rheological behaviour comparable to the binder initially recovered from RA. The re-ageing was investigated by Li et al. [51] and Zhou et al. [55], to respectively assess the effectiveness of warm mix asphalt (WMA) additive (5% by the total weight of the virgin and aged binders) and the anionic styrene-butadiene rubber (SBR) latex (4% by aged binder weight) used in combination with ARA (4% by aged binder weight), and the use of a virgin SBS-modified binder in place of a neat one. The results showed that adding WMA agent, SBR latex and using SBS-modified binder play a positive role in fatigue and low temperature cracking performance after re-ageing. According to the results by Caputo et al. [50], all ARAs were capable of recovering the physical properties to varying extents. However, not all of them could restore the chemical features, as found using different chemical testing analyses. Epps et al. [38], as a part of the large USA national study (NCHRP Project 09–58) on the impact of different ARAs on high RAb and mixtures, showed that both the type and chemistry of the ARAs and the quality and properties of the virgin bitumen had a significant impact on the ageing performance of high RA mixtures. Furthermore, through evaluation of the impact of multiple cycles of RTFOT and PAV ageing on various rheological and Black space parameters, they showed that the use of single threshold values for such parameters may not be sufficient, as the rate of change of such parameters with continued ageing may be a significant differentiating factor for RA mixtures containing ARAs.

The dissimilar findings of numerous studies were related to the effectiveness of the rejuvenation process, particularly the type and dosage of ARAs [56, 57]. The proper ARAs selection criteria to create a relatively homogeneous RAb are the ability to lower the viscosity of an aged binder, reduced mass loss due to heating, and a higher flashpoint. ARAs might be incompatible with asphalt or diffuse slower, forming a nonhomogeneous Rab. Generally, the long-term interaction between ARAs and RAb is still an open task.

Given the current state of the knowledge, it appears that one of the lacking aspects related to the previous research efforts is the possibility of consistently comparing the available results. This is more often associated with the diverse research strategy and experimental procedure adopted in past and recent investigations. Therefore, there is an undeniable need for approaching the research on rejuvenators under a single framework and scientific protocol capable of providing solid support to the results. This is the cornerstone of the research work developed and performed by Task Group 3 on Asphalt Binder for Recycled Asphalt Mixtures.

4.3 Objective and Research Approach

A joint research effort, initially including 20 laboratories from different locations in the world, is a significant challenge in terms of logistics (for example, sample shipment), meetings, and consistency of the testing methods. That being said, those interlaboratory activities produce beneficial results that are difficult to obtain with a single lab and are frequently carried out in RILEM.

A hybrid approach between a conventional interlaboratory activity and more open experimentation was adopted to study the ageing behaviour of ARA binder blends. This had the purpose of facilitating the comparison of results on a core set of materials while limiting the delivery of the material to a reasonable level to the participants. Because of the complexity of organizing such a vast research effort, the overall activity was planned into three steps. In the beginning, the same material and test methods were used in the interlaboratory activities, while in the further steps, additional materials and experimentation were included.

In order to study the ageing behaviour of the ARAs, a set of binder blends was produced from an extracted and recovered RA binder (RAb), Vb, and an ARA. These blends were subsequently subjected to different ageing levels (STA and LTA). All the blends were tested, and the changes in conventional parameters, rheological behaviour and chemical characteristics were analysed. The asphalt mixture used three different contents of RAb to simulate a different recycling content ($R_c = 60\%$, 80% and 100% by weight).

Regarding the limited laboratory resources, a three-step research approach was initially agreed upon to study the ageing behaviour of ARA binder blends:

Step 1 A core experiment with the same material and the same test methods was used to "calibrate" the different laboratories and assess the influence of the slight differences in the similar test methods of ASTM and CEN standards. With three different simulated RA contents, R_c, of $60R_c$, $80R_c$ and $100R_c$, seven laboratories consistently shared the same material composition producing the same number of results, which were considered satisfactory to provide a robust benchmark for all participants.

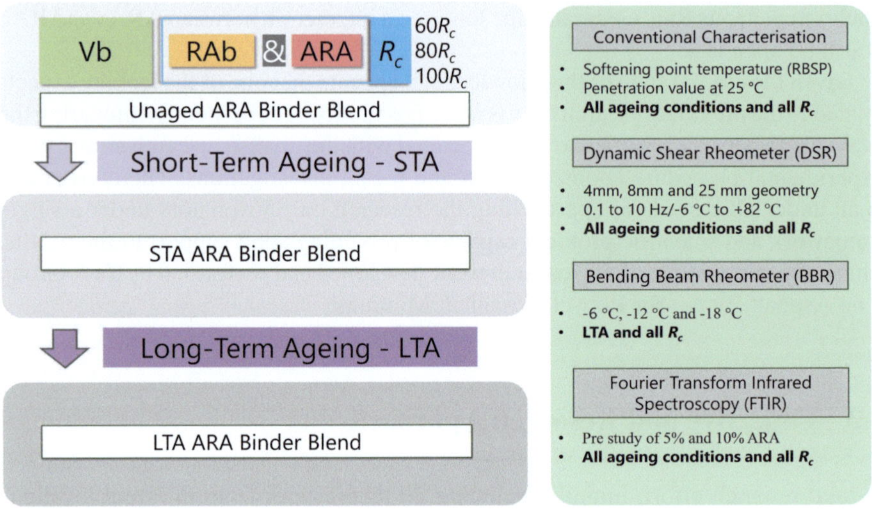

Fig. 4.1 Schematic of the interlaboratory testing

Step 2 An RA binder extracted by each laboratory from a local RA source was allowed to replace the core RAb. The same ARA and Vb adopted in *Step 1* were used to limit the change to only one parameter. This enabled delivering information on the impact of the RA source and properties.

Step 3 In this step, participating laboratories were allowed to use other virgin binders and/or ARA together with the above national RAb (from *Step 2*), with the restriction to introduce only one additional new parameter in the analysis.

This three-step approach was applied to guarantee that all the results were comparable with the results of the initial core experiment. The test procedures and ageing methods were identical in all three steps. The schematic of the interlaboratory testing methodology, including a list of the tests described in Sect. 5, is illustrated in Fig. 4.1. This applies to all the three steps (research approach) described above.

4.4 Material and Experimental Methods

Section 4.4 presents the general experimental approach adopted in the present research. In the following Sect. 5, the different materials used, and the test methods selected for the interlaboratory activities are briefly described.

4.4.1 Materials

Different materials were used depending on the three different research steps previously described in Sect. 4.4. In *Step 1*, the core material consisted of three components: Vb, RAb, and ARA. The selected Vb was an unmodified penetration graded [58, 59] 50/70 binder. This material was obtained from a single source and a single batch from a large binder producer. The Technical University of Braunschweig (Germany) (Lab 18) collected the RA material from a local provider in the north of Germany and performed the extraction and recovery process of the RAb [60]. An initial characterisation of these two materials, including the Performance Grade (PG) [61], is reported in Table 4.1. In addition to conventional testing, the potential presence in the RAb of polymers or residual solvent from the extraction and recovery process was investigated with Fourier Transform Infrared Spectroscopy (FTIR) in Attenuated Total Reflection (ATR) mode showing that neither of these two materials was contained in the extracted RAb.

The ARA used in the current research consisted of a bio-based additive. This liquid-like material presents an amphipathic chemical structure capable of dispersing the highly polar fractions, which minimize the agglomeration of asphaltenes [34]. Such material has already found significant applications and has been evaluated in different research and practical projects [62, 63]. Some of the properties of the ARA are presented in Table 4.2.

Three different ARA-RAb-Vb blends simulating the recycling contents, R_c, of $60R_c$, $80R_c$ and $100R_c$ were produced according to Table 4.3. The blends were prepared by first heating the 50/70 Vb and RAb at 130 °C in the oven to soften the material. After selecting the desired amount of RAb, the design content of (non-heated) ARA by weight was added to the RAb and stirred for 30 s. This mix of ARA and RAb (ARA-RAb) was placed back at 130 °C in the oven for 10 min and stirred for an additional 30 s at 5 min intervals for homogenization purposes. Then, the design amount of Vb was combined with the ARA-RAb mix and stirred for 30 s. Similarly to the previous procedure, the ARA-RAb-Vb blend was placed in the oven at 130 °C for 10 min and stirred for 30 s at 5 min intervals to complete the homogenization of the blend. This research used a similar procedure for all three steps (1, 2, and 3). It

Table 4.1 Initial characterisation of virgin and RA binder

Binder	Penetration at 25 °C (0.1 mm)	Softening point (°C)	Penetration index	PG
Vb 50/70	62	49.6	−0.8	70–22
RAb	14	71.0	0.3	88–10

Table 4.2 ARA: properties

Flashpoint	Viscosity at 60 °C	Cloud point	Density
> 280 °C	22 10^{-3} m^2/s	<−25 °C	0.93 Mg/m^3

Table 4.3 Blending Table (% = percentage by weight)

Blend component	Recycling content—R_c		
Content	60%	80%	100%
ID	60R_c	80R_c	100R_c
ARA (%)	5.0	6.6	8.3
RAb (%)	55.0	73.4	91.7
ARA/RAb (%)		9.0	
Vb (%)	40.0	20.0	0
Tb (%)	100	100	100

R_cRecycling content = ARA + RAb,
ARAAsphalt Recycling Agent.
RAbReclaimed Asphalt binder content– Equivalent binder replacement ratio.
VbVirgin binder content,
TbTotal binder content = Vb + RAb + ARA = 100%,

must be noted that the blending of Vb and ARA-RAb was not performed when the 100% recycling level was the specified design.

A constant ARA-RAb ratio of 9%, determined by a pre-test, was used to soften the RAb to the same needle penetration [64] value as the original 50/70 Vb (Table 4.1). Hence, the penetration value should be equal for all ARA-RAb-Vb and ARA-RAb blends, just after their preparation (unaged condition). Penetration was chosen as the target value; however, other options, such as softening point temperature [65, 66] and shear modulus $|G^*|$ [67, 68], were considered.

In *Step 2* of the research, the participating laboratories were free to select a local RA to be used. Only two laboratories could commit to using a different RA material.

In *Step 3*, different Vb and ARA could be selected for the experimental campaign. Similar to step 2, only 3 of the laboratories could conduct the experimentation.

The ARA-RAb-Vb blends were subjected to ageing procedures to consider the effect of ageing. STA was carried out with the Rolling Thin Film Oven Test (RTFOT) [30, 32], and LTA with the Pressure Aging Vessel (PAV) [31, 33].

4.4.2 Experimental Methods

The testing campaign consisted of several tests, displayed in Table 4.4, together with corresponding standards. Given such a large number of laboratories, the specifically adopted testing methods resulted from a compromise between the capability of the different testing facilities and the limited amount of recovered RAb to cover a sufficiently diversified testing program across the laboratories.

Table 4.4 Testing methods and appropriate standards used in the study

Testing method	Standard(s)
Softening point temperature (RBSP)	EN 1427 [65], ASTM D36/D36M-14 [66]
Penetration value at 25 °C (Pen)	EN 1426 [59], ASTM D5/D5M-20 [58]
Linear viscoelastic properties—Dynamic Shear Rheometer (DSR)	EN 14,770 [67], ASTM D7175-15 [69]
Low temperature characterization—Bending Beam Rheometer (BBR)	EN 14,771 [70], ASTM D6648-08 [71]
Chemical characterization—Fourier Transform Infrared Spectroscopy (FTIR)	Attenuated Total Reflection mode [72]

4.4.2.1 Conventional Characterisation

The Technical University of Braunschweig provided the RAb to all participating laboratories. In this way, no additional uncertainty was introduced by the RAb extraction/recovery process while also avoiding the complexity of shipping large amounts of RA mixture to the laboratories.

Initial characterisation of Vb, RAb, and ARA-RAb-Vb blends was performed with penetration value and softening point methods. Both penetration values at 25 °C and softening point temperature were measured on the materials at the different ageing conditions.

Pen and BBR tests were carried out only by some laboratories, as the amount of RAb was insufficient for all laboratories to complete an entire testing program. Also, laboratories were free to perform additional testing not included in the core experimental program synthetically presented in Table 4.5. This includes, for example, FTIR, among others. A total of 17 laboratories were able to provide experimental data for the core material. In addition, Lab 12 and Lab 20 performed a series of DSR and BBR tests on different RAb sources.

4.4.2.2 Rheological Tests

The viscoelastic behaviour of different materials was assessed using the DSR [67–69]. Three different geometry plates (25 mm, 8 mm and 4 mm) were used for three different temperature ranges to perform temperature and frequency sweep tests. The recommended test frequencies were: 0.1, 0.2, 0.3, 0.4, 0.5, 0.6, 0.7, 0.8, 1, 1.59, 2, 3, 4, 5, 6, 7, 8, 9, and 10 Hz for each set of plate geometries. The 25 mm plate geometry with a 1 mm gap was adopted for temperatures between 28 °C and 82 °C with an incremental step of 6 °C. The 8 mm plate-plate geometry and 2 mm gap were selected at the intermediate temperature range (T = −6, 0, 4, 10, 16, 22, 28, 34, 40 °C). The 4 mm plate geometry with a 1.75 mm gap was used at the low temperatures with suggested temperatures of -30, -24, -18, -12, −6, 0, 4, 10 °C [73, 74]. The latter

Table 4.5 Overview of the interlaboratory testing matrix for the ARA-RAb-Vb blends

Lab No	Country	Testing methods					Recycling content - R_c			Vb	RAb
		Pen	RBSP	DSR	BBR	FTIR	$60R_c$	$80R_c$	$100R_c$		
1	JPN	√	√	√					√		√
2	AUT	√	√	√			√				
3	ITA		√	√					√	√	√
4	USA			√			√				
5	USA		√				√			√	
6	USA	√	√	√			√				
7	NLD	√	√	√	√	√		√		√	√
8	PRT	√	√	√		√	√				
9	BEL	√	√	√		√		√			
10	CHE		√	√				√		√	
12	ITA		√	√			√				
14	CAN		√	√					√		
15	BRA			√					√		
16	FRA	√	√	√	√		√				
18	GER	√	√	√	√			√		√	
19	SRB	√	√	√				√			
20	CZE	√	√	√	√				√		√

geometry was used only by a limited number of laboratories equipped with such a testing setup.

Tests could be conducted in stress or strain control depending on the specific testing device available or in a mixed-mode (stress control for the lower temperatures and strain control for the higher temperatures). Amplitude sweep tests preceded temperature and frequency sweep (T-f-sweep) tests to determine the required testing stress and strain to be used in the actual DSR tests to remain in the linear viscoelastic domain. The results of the DSR testing were used to obtain a broad spectrum of complex shear modulus ($|G^*|$) and phase angle, δ, together with the high PG [61].

4.4.2.3 BBR

BBR [70, 71] tests were performed after combining STA and LTA at the following temperatures: $T = -6, -12, -18$ °C. A minimum of two replicates per test temperature was used. Creep stiffness $S(t)$ and m-value were recorded throughout the test. The critical temperatures, the low PG [61], and the continuous low PG [75, 76] were then calculated and compared. It should also be mentioned that due to the limited amount of available material, four laboratories ran the BBR tests for seven different binders with different recycling content and sources, respectively.

4.4.2.4 FTIR

FTIR test was conducted by three laboratories. ATR was used in all cases. FTIR profiles were then analysed and results were compared across the different materials.

4.5 Results, Analysis and Discussion

Section 6 presents the most significant results obtained during the activity of Task Group 3 for conciseness. Mainly focusing on the results from *Step 1* and part of *Step 2*, this includes the initial conventional characterisation, a selected set of DSR and BBR output, and a synthesis of the FTIR results,

4.5.1 Initial Conventional Characterisation

Among the 20 laboratories, 16 performed RBSP measurements on the different bituminous blends, and ten conducted pen tests at 25 °C. In addition, two laboratories performed both measurements on the Vb while three laboratories conducted the conventional characterizations also on RAb.

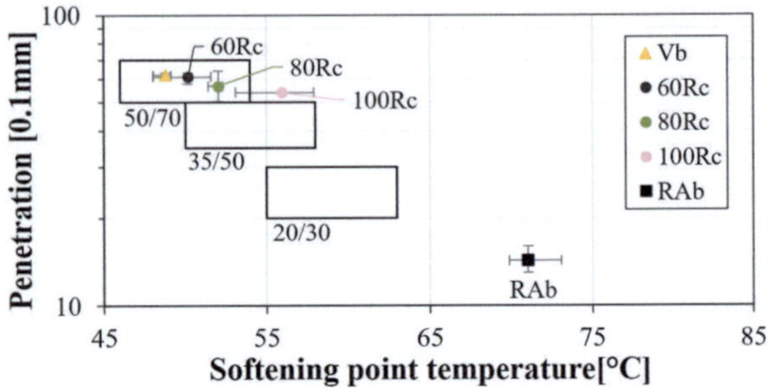

Fig. 4.2 Penetration value and softening point temperature for unaged binders

Figure 4.2 displays the basic properties of the Vb and the RAb and the blends at their unaged state. The error bars report the min and max values from all laboratories. Four laboratories reported values for RAb content of $60R_c$ and $80R_c$ and two for RAb content of $100R_c$. The RAb was relatively hard, with a penetration value at 25 °C of 16 0.1 mm. In the case of the blends, the penetration values were consistently restored regardless of the RAb content. Concerning the RBSP temperature, the values were more influenced by the amount of RAb in the blend. Although for the $100R_c$, two laboratories recorded values: one was within the specification of a 50/70 penetration range, while the other reported a higher softening point. The dosage of ARA enabled restoring the intermediate temperature properties while keeping the positive effect of the aged binder on the high temperature characteristics.

Figure 4.3 reports the changes over ageing for the different types of bituminous materials and blends as an average from all laboratories. Open, shaded and solid markers represent the unaged, RTFOT and RTFOT + PAV aged material. Concerning the Vb, the properties changed by one pen-grade harder after each ageing state. The RAb became harder over ageing but with a reduced degree of change than the Vb associated with an increase in RBSP of 12 °C and 14 °C after combined STA and LTA. In the case of three RAb content at $60R_c$, $80R_c$ and $100R_c$, the changes followed similar trends with a more minor change in properties across ageing compared to the 50/70 Vb. Under the assumption that the ARA would vanish after RTFOT and RTFOT + PAV ageing, the penetration and RBSP temperature values would have been closer to those from the RAb.

In *Step 2* of the interlaboratory testing, laboratories were able to substitute the core *Step 1* RA with a local RA. Two laboratories investigated the ageing behaviour of the ARA-RAb-Vb blends according to the procedure of *Step 1* and provided RBSP (Fig. 4.4) and DSR results (see subsection 6.2). Laboratory 20 used a slightly softer RA (identified as CZE RAb and presenting a softening point equal to 63.0 °C) in combination with lower ARA content. The evolution of the ageing was identical to Vb and exhibited a similar trend to that of the blends prepared with the core *Step 1* RAb. However, this was not the case for Lab 12 as the RAb used in the RILEM TC 264-RAP TG2, on Non-Cold (Warm and Hot) Recycling (hereafter identified as TG2 RAb), was adopted for the ARA-RAb-Vb blend produced for *Step 2*. The increase in

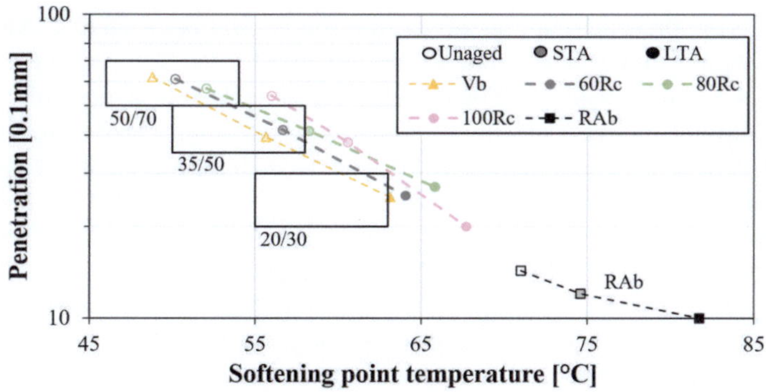

Fig. 4.3 Penetration value and softening point temperature of binders over ageing

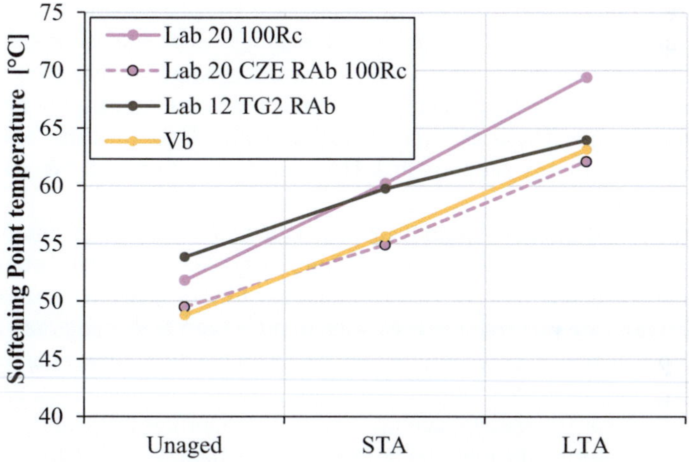

Fig. 4.4 Softening point for ARA-RAb-Vb blends with different RAb—*Step* 2

RBSP was significantly lower, especially after LTA, compared to the other blends. This shows that the RA binder might influence rejuvenated blends' ageing behaviour.

4.5.2 DSR Results and Analysis

This subsection addresses the combined effect of RAb content and ageing conditions on the rheological properties of ARA-RAb-Vb blends. For this purpose, Black diagram and Cole–Cole plots are selected to present and compare the results visually. The specific graphical representation enables displaying the raw data at all frequencies and temperatures without any shifting and optimization process. Some of the results are presented in Fig. 4.5 as an example. In Figs. 4.5a to c, ARA-RAb-Vb blends with different RAb contents are reported for the same ageing condition, while Figs. 4.5d and e display the blends with the same RAb contents under different ageing conditions. As illustrated in Fig. 4.5a and c (black diagram of unaged and STA ARA-RAb-Vb blends), no remarkable differences can be observed between different RAb contents, except in the case of Lab 3. Overall, it appears that for the specific ageing condition, similar rheological response can be obtained for the different RAb contents with a closer trend for higher degrees of ageing. In Fig. 4.5b, the Cole–Cole plot of the ARA-RAb-Vb blends in the unaged condition is provided as an example. Due to the specific characteristics of this type of plot and the relatively high testing temperature (above −6 °C), it is quite difficult to distinguish the differences between each laboratory. In the Cole–Cole plot, data obtained at temperatures above zero are not clearly visible as they are squeezed toward the origin [77]. Hence, only black diagrams are used later in the analysis. In Figs. 4.5d and e, both types of RAb content blends show similar trends in all ageing conditions. The phase angle decreases while the shear modulus increases when the materials age. It should be noticed that in both RAb contents, narrow curves are found within a single ageing group. Overall, Fig. 4.5 suggests a robust reproducibility of the data obtained during the interlaboratory experimentation. The consistency of the curve trends for the same material and ageing condition visually supports this. Hence, the average results are used for the following analysis.

In addition to the black-diagram representation, the average of complex shear modulus, $|G^*|$, and phase angle, δ, are plotted against the testing temperature at the conventional frequency of 1.59 Hz (10 rad/s) for further analysis. The isochronal plots of G^* and δ are presented in Figs. 4.6a to d for both *Step 1* and *Step 2*, in both steps, the Vb is displayed as the reference material for the different ageing conditions. In Fig. 4.6a and b, it is found that there are no remarkable differences as the RAb content changes for the specific ageing condition supporting the idea that the ARA is consistently restoring the material properties. Nonetheless, it visually appears that the binder blends under unaged conditions seem to show similar rheological values with short-term aged material. This is especially true for modulus results, whereas a more distinct trend across ageing levels can be observed in the case of phase angle. In the case of Fig. 4.6c and d, the ARA-RAb-Vb blends prepared with the alternative RAb sources used in *Step 2* present similar rheological properties with the reference materials both in terms of G^* and δ. However, a slightly lower G^* and higher δ are observed than the RAb used in *Step 1*. This suggests, as previously observed for the conventional characterization, that the RAb source has a remarkable influence on the rheological properties of the blends.

Together with a simple graphical representation of the results, the high temperature Performance Grade (PG) [61] and LTA temperature criteria were calculated for the different laboratories (Table 4.6). For most of the laboratories, a high PG limit of 70 was obtained independently of the different R_c and Vb material. Three laboratories (Lab 2, Lab 3, and Lab 6) reported different high PG of 64 (highlighted in bold in Table 4.6). However, a deeper observation revealed that the continuous high PG of the three laboratories was very close to 70, especially in the case of Lab 2 and Lab

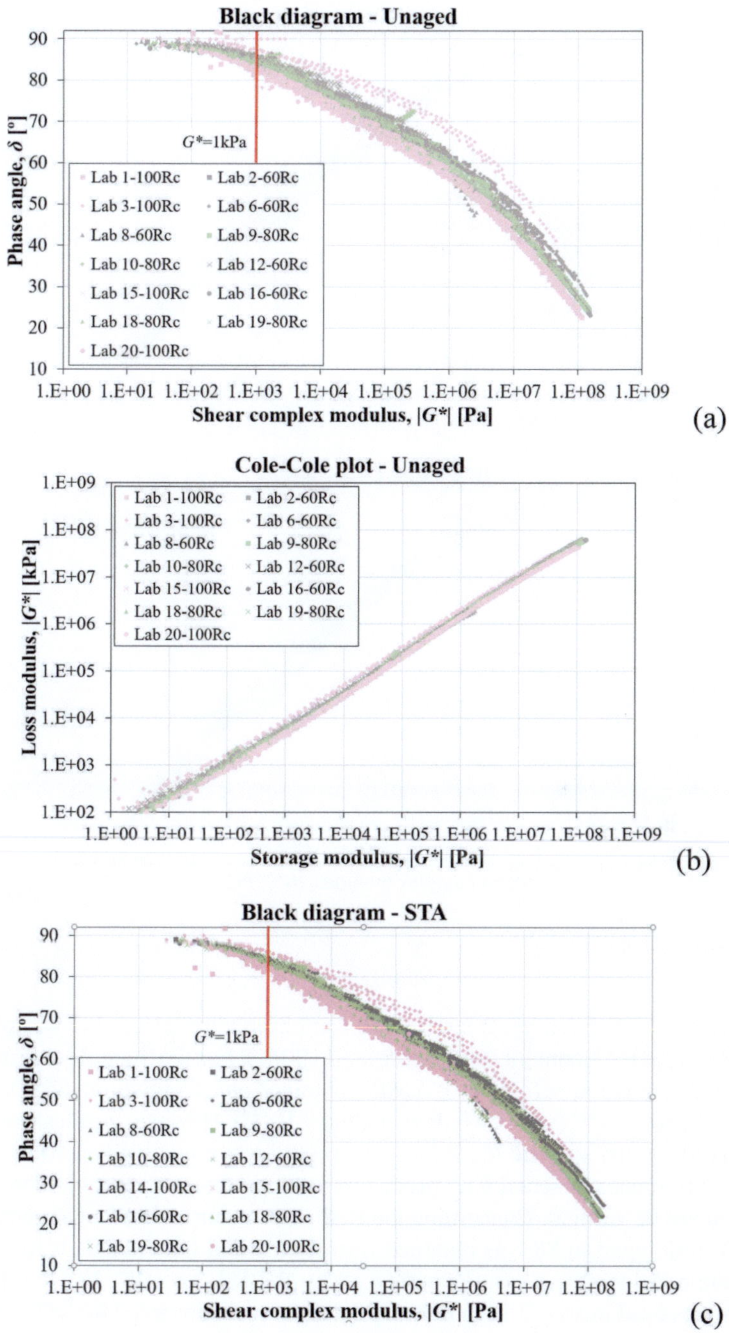

Fig. 4.5 Evaluation of repeatability of ARA-RAb-Vb blends under different ageing conditions and RAb content: **a** black diagram - unaged; **b** Cole–Cole plot - unaged; **c** black diagram - STA and **d** black diagram—$60R_c$; and **e** black diagram - $80R_c$

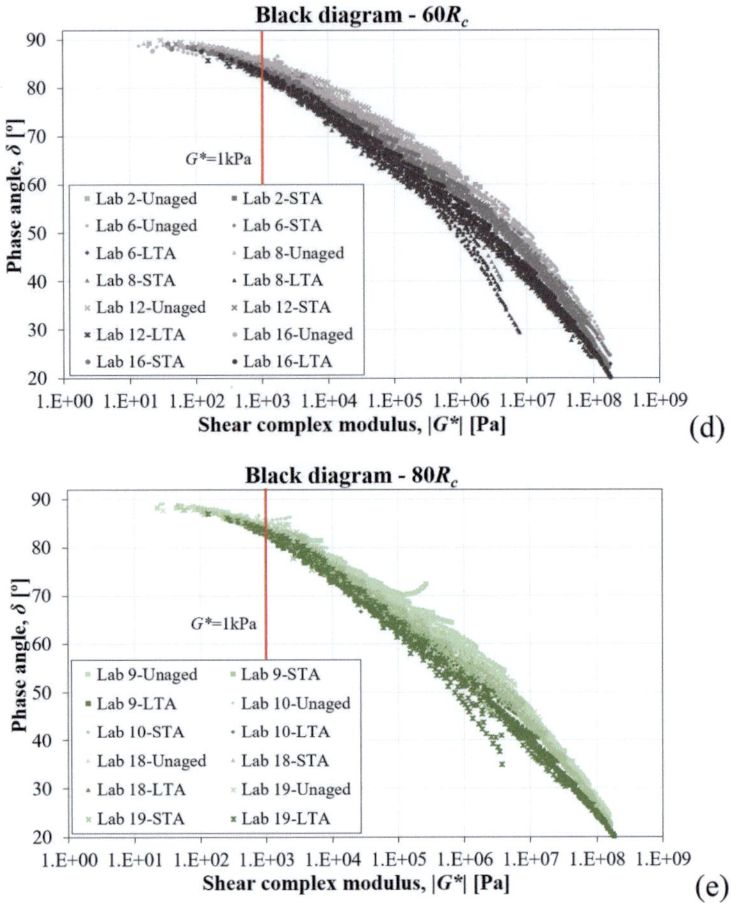

Fig. 4.5 (continued)

6 (less than 1 °C difference). Overall, consistent high temperature PG was obtained for the different RAb content (60R_c, 80R_c and 100R_c). For the Superpave criterion at intermediate temperature ($G^*\sin\delta \leq 5$ MPa) obtained on LTA material, $G^*\sin\delta$ ranges from 15.8 °C to 24.4 °C (with a window of almost 10 °C). However, no clear trends can be observed for any specific R_c. While this might be hypothetically associated with possible differences in actual LTA procedures among the participating laboratories, this could not be verified. Concerning the RAb, it is unsurprising that a much higher high PG limit equal to 88 was obtained from Lab 3 and Lab 7. This is most likely due to the considerable ageing experienced by the RA material. For the different RAb sources used in *Step 2*, high PGs of 70 and 64 were observed for TG2 and CZE RAb, respectively. It should be noticed that CZE RAb presented an actual continuous high PG = 68.0, which is relatively close to the next PG level of 70. These findings

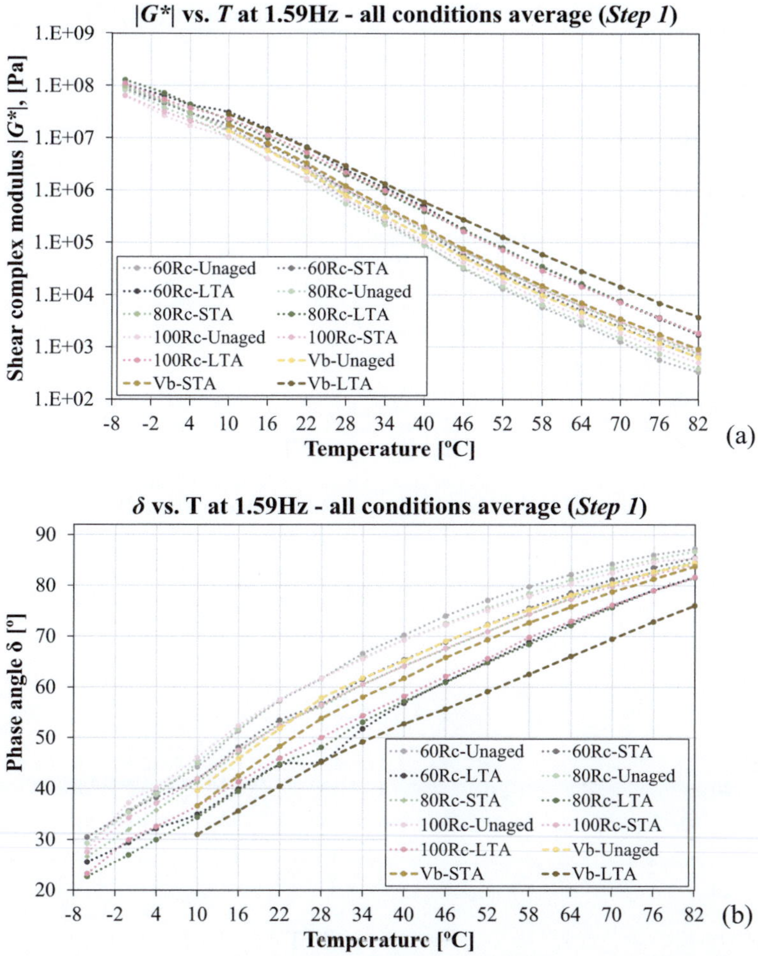

Fig. 4.6 Average isochronal rheological parameter under different temperatures at 1.59 Hz (10 rad/s): **a** G^* and **b** δ

are consistent with the initial characterisation, including penetration value at the temperature of 25 °C and RBSP temperature.

4.5.3 BBR Results and Analysis

The raw experimental results of BBR in terms of creep stiffness, $S(t)$, and relaxation parameter, m-value(t), at $t = 60$ s are shown in Fig. 4.7. Very similar $S(t)$ and m-value(t) can be observed for $60R_c$ and $80R_c$. Additionally, a limited increase in creep

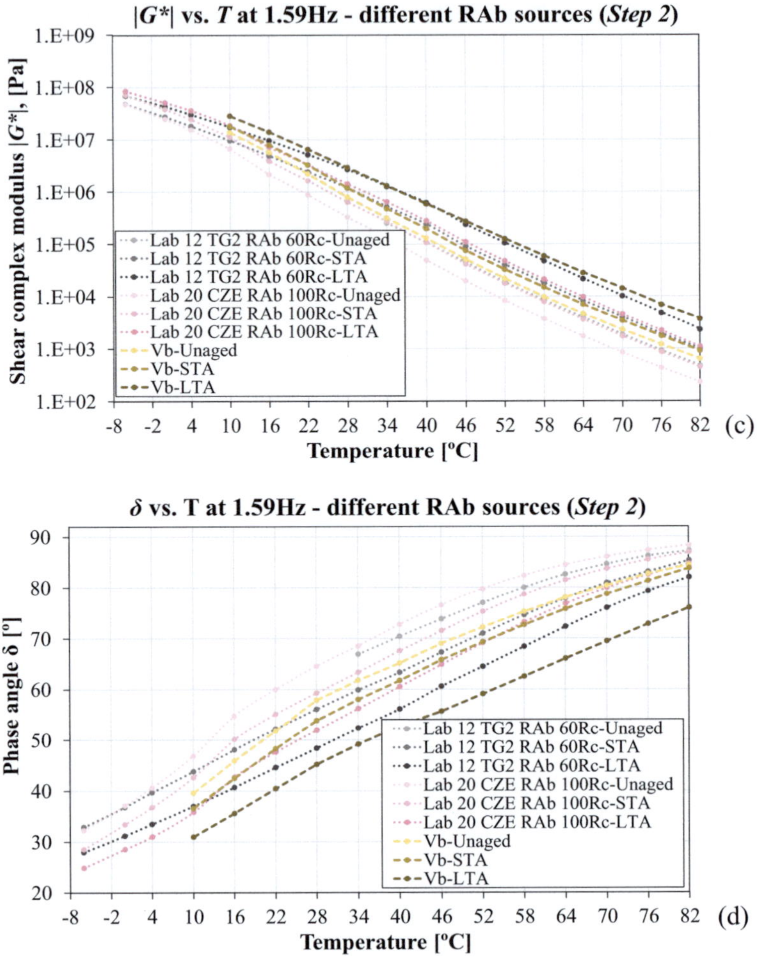

Fig. 4.6 (continued)

stiffness and reduction in *m*-value were found. The continuous critical temperatures were calculated based on stiffness, $T_c(S)$, and *m*-value, $T_c(m)$, criteria and listed in Table 4.7. It can be noted that with the increase in RAb content, $T_c(S)$ gets slightly lower. In the case of $T_c(m)$, similar values were found for $60R_c$ and $80R_c$, while a remarkable increase was achieved by $100R_c$. The related continuous PG and real low PG limits were also computed (Table 4.7). The same low PG limit was found in the case of $60R_c$ and $80R_c$, whereas a higher level was observed for $100R_c$. It can be noticed that the use of ARA can soften the aged RAb; however, only part of the relaxation properties could be restored. When an alternative RAb source was used to design an ARA-RAb-Vb with $100R_c$ (CZE RAb), as in the case of Lab 20, a low PG of −28 was observed. According to the results of Table 4.6, a lower high PG

Table 4.6 DSR Performance Grade (PG) criteria for different R_c

Materials	Lab No	unaged (°C) ($G^*/\sin\delta \geq$ 1.0 kPa)	STA (°C) ($G^*/\sin\delta \geq$ 2.2 kPa)	LTA (°C) ($G^*\sin\delta \leq 5\ MPa$)	High PG (°C)
Vb	3	77.6 (70)	74.6 (70)	20.4	70
	7	71.5 (70)	71.3 (70)	24.2	70
RAb	3	92.5 (88)	–	–	88
	7	94.3 (94)	92.1 (88)	30.8	88
60R_c	2	70.6 (70)	69.8 (64)	-	64
	6	69.1 (64)	69.6 (64)	20.0	64
	8	70.2 (70)	70.4 (70)	24.4	70
	12	72.1 (70)	72.0 (70)	16.1	70
	16	71.1 (70)	70.9 (70)	19.6	70
80R_c	7	75.5 (70)	74.6 (70)	21.0	70
	9	71.2 (70)	71.5 (70)	18.0	70
	10	73.7 (70)	72.2 (70)	19.7	70
	18	71.7 (70)	70.6 (70)	16.4	70
	19	73.1 (70)	72.9 (70)	16.5	70
100R_c	1	78.9 (70)	74.3 (70)	19.8	70
	3	69.8 (64)	68.1 (64)	21.7	64
	14	-	71.8 (70)	18.5	70
	15	74.6 (70)	72.3 (70)	15.8	70
	20	75.2 (70)	74.3 (70)	19.6	70
60R_c TG2 RAb	12	-	74.3 (70)	18.5	70
100R_c CZE RAb	20	68.5 (64)	68.0 (64)	15.8	64

limit was achieved for this material. This further confirms that the RAb source can significantly influence the rheological properties of the asphalt binder.

4.5.4 FTIR Results and Analysis

FTIR has already been successfully used in the previous RILEM TCs to assess the ageing of bituminous binders, such as in RILEM TC 206-ATB [9] or RILEM TC 252-CMB [11]. Over ageing, oxygen intake can be identified and tracked through the increase of carbonyl, C = O, around 1700 cm^{-1}, and sulfoxide S = O species, around 1030 cm^{-1}. Furthermore, it enables the detection of any additional chemical functional groups from modifiers when their fingerprint can be distinguished

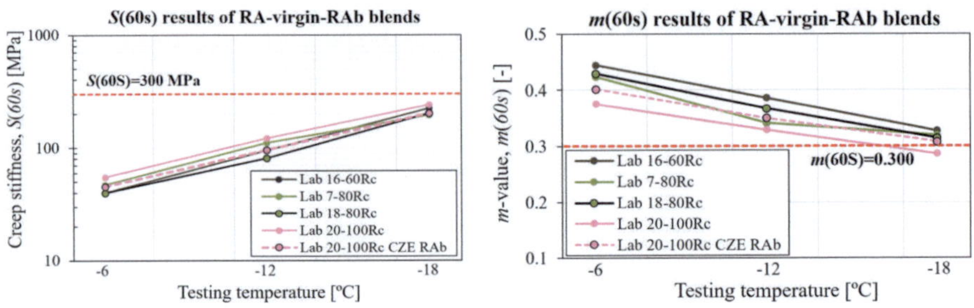

Fig. 4.7 BBR experimental results for long-term aged ARA-RAb-Vb blends: (left) creep stiffness; (right) m-value

Table 4.7 BBR calculated critical continuous temperatures and real PG limit

RAb content	Lab No	$T_c(S)$ [°C]	$T_c(m)$ [°C]	ΔT_c [°C]	Continuous Low PG	Low PG
Vb	7	−14.7	−16.0	1.3	−24.7	−22
RAb		−9.2	−3.1	−6.1	−13.1	−10
$60R_c$	16	−20.0	−20.7	0.7	−30.0	−28
$80R_c$	7	−21.1	−19.0	−2.1	−29.0	−28
	18	−21.5	−19.5	−2.0	−29.5	−28
$100R_c$	20	−19.6	−16.1	−3.5	−26.1	−22
$100R_c$ CZE RAb	20	−21.2	−18.8	−2.4	−28.8	−28

from the conventional binder fingerprint. For example, this is the case with Styrene–Butadiene–Styrene, SBS block co-polymer, as reported by Mouillet et al. [78]. Most often, FTIR is performed in ATR.

FTIR was first used during the pre-dosage study. It was performed on the RAb to identify the possible presence of a polymer or residual solvent from extraction. None of them was detected. It was also conducted on the ARA alone to identify its typical fingerprint, and later the blends at 5% and 10% dosage on RAb. Figure 4.8 shows the FTIR spectrum for the pre-dosage study between 1800 and 600 cm^{-1} wavenumbers, for the RAb, the ARA, and the blends with 5% and 10% of ARA.

The peaks for styrene or butadiene, typically around 750 cm^{-1} or 965 cm^{-1}, could not be identified. This meant that the RAb most likely did not contain SBS polymer. The ARA displayed clear characteristic peaks, which can be isolated from the standard binder spectrum with a sharp peak at 1740 cm^{-1} (ester) and a broader peak around 1150 cm^{-1}. These peaks could be seen in the blends proportionally to the amount of additive. The sharper the peak, the easier it is to identify and track it.

During the interlaboratory testing, Lab 7, Lab 8 and Lab 9 performed FTIR, all in ATR mode. Tests were performed on ARA, RAb and Vb, with one laboratory having also performed measurements on binder blends after STA and LTA. For the additive,

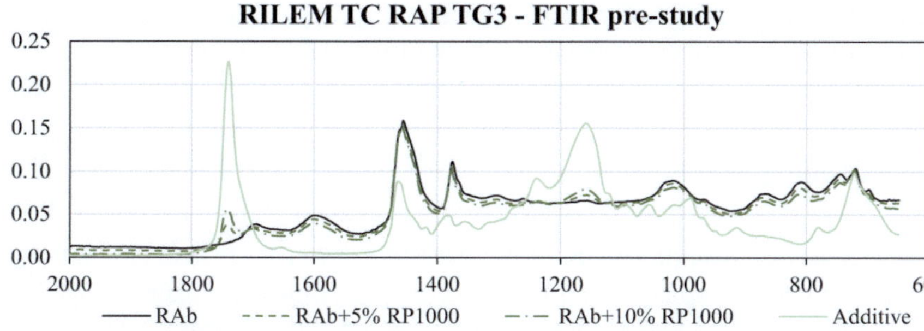

Fig. 4.8 FTIR on RA and blends incorporating 5% and 10% ARA form the pre-study

the sharp peak around 1740 cm^{-1}, corresponding to the ester group, is consistent and can be used to track the presence of the additive.

Figure 4.9 displays the results of the blend with 80R_c RA from Lab 7. The sharp peak at 1740 cm^{-1} from the additive can be seen and remained consistent at the three different ageing conditions. The carbonyl and sulfoxide peaks are mostly from the RAb and are unaffected by the ARA. This means no chemical interaction with the binders and no additional increase in the different peaks over ageing is experienced [79]. The sharp peak from the additive is close to the carbonyl peak, and they may overlap and need special attention in interpretation.

FTIR can be further used to calculate indexes for carbonyl and sulfoxide as the ratio of the respective area over aliphatic peak areas, which are not affected by ageing. The way to determine these indexes, especially with area boundaries, may directly impact the value. In RILEM TC206 ATB [9], for Carbonyl at 1700 cm^{-1}, the boundaries were suggested in the range of 1660 cm^{-1} and 1753 cm^{-1}. These boundaries are broad and may be valid in the case of the RAb. However, they also include the peak from the ARA at 1740 cm^{-1}. Thus, it is essential to refine the boundaries to

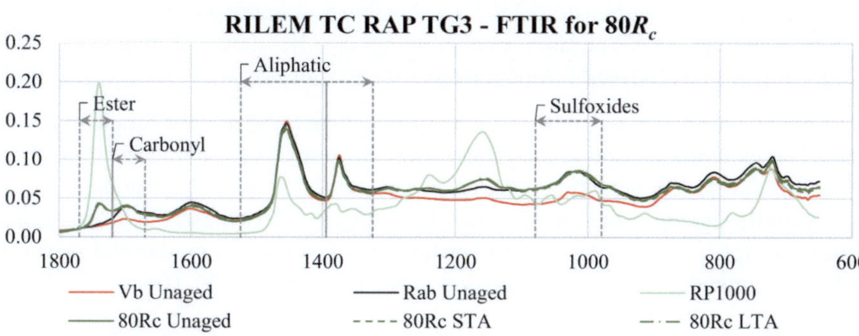

Fig. 4.9 FTIR on ARA-RAb-Vb blend (80R_c) over ageing from a single laboratory

specifically distinguish between carbonyl and ester groups from the ARA, with the ester from 1720 to 1770 cm^{-1} and the carbonyl from 1670 to 1720 cm^{-1}.

4.6 Conclusions

In this chapter, the overall activity of the RILEM TC 264-RAP Task Group 3 was presented. A research approach has been proposed to evaluate the ageing behaviour of binder blends prepared with Vb, RAb and an ARA. For this purpose, conventional, rheological, and chemical tests were used. Based on the research results, the following conclusions can be drawn:

- The proposed research approach was capable of comprehensively address the effect of the selected ARA on the ageing behaviour of bituminous blends containing different proportions of aged bitumen from reclaimed asphalt;
- both the conventional characterization and the rheological analysis performed with the dynamic shear rheometer support the idea that ARA was capable of substantially restoring the rheological properties of the ARA-RAb-Vb blends with a response similar to those of the original Vb selected as reference material;
- the ageing effect on the ARA-RAb-Vb blends is comparable to that experienced by the reference Vb with similar trends for an increasing degree of ageing. This is also confirmed by conventional characterization and rheological measurements;
- at the lower temperatures, the limited available data suggest that the effect of the ARA is more diversified. Specifically, in the case of the highest recycling content ($100R_c$) of RAb, less effective restoration of the relaxation properties is experienced. This is reflected in the difference in critical temperature and performance grade across the three different RAb contents;
- when different RAb sources were used for preparing ARA-RAb-Vb blends, different trends in the empirical and rheological response were observed depending on the selected RAb and the rejuvenating effect of the ARA;
- the FTIR analysis demonstrated to be a valuable tool in the pre-dosage phase and to evaluate the ARA's possible impact on ageing. No chemical interaction with the binders was observed for the different degrees of ageing;

Although based on a sizeable interlaboratory activity, the present conclusions are centred on experimentation at the binder level, specifically on unmodified binders. To confirm further the potential of the proposed methodology and extend to research activity on the ageing behaviour of ARA, it is suggested to expand the investigation to more complex bituminous material phasees such as mastic, mortar and mixture. Such a research effort could be addressed in a follow-up RILEM TC.

Acknowledgements The RILEM Technical Committee on Asphalt Pavement Recycling (TC 264-RAP) and Task Group 3 Members are gratefully acknowledged. The authors would also like to thank Nynas AB and Kraton Chemical B.V. for supporting the interlaboratory activity by providing bituminous materials and ARA. The authors are grateful for the support provided by following collaborators:

• Aurelie Lorserie (Campus Scientifique et Technique (CST) COLAS, 4 Rue Jean Mermoz, 78,771 Magny-les-Hameaux, France)

• Goran Mladenović (University of Belgrade, Faculty of Civil Engineering, Bulevar kralja Aleksandra 73, Belgrade, Serbia)

• Runhua Zhang (Department of Civil and Environmental Engineering, University of New Hampshire, 33 Academic Way, Durham, NH 03,824, US)

• Tomas Koudelka (Faculty of Civil Engineering, Institute of Road Structures, Brno University of Technology, Brno, Czech Republic)

• Pavel Coufalik (Faculty of Civil Engineering, Institute of Road Structures, Brno University of Technology, Brno, Czech Republic).

References

1. Du, Z., Zhu, X., Li, F., Zhou, S., Dai, Z.: Failure of the asphalt-aggregate interface under tensile stress: Insight from molecular dynamics. J Mater Civil Eng **33**(3), 04021008 (2021). https://doi.org/10.1061/(ASCE)MT.1943-5533.0003601
2. Kriz, P., Grant, D.L., Veloza, B.A., Gale, M.J., Blahey, A.G., Brownie, J.H., Shirts, R.D., Maccarrone, S.: Blending and diffusion of reclaimed asphalt pavement and virgin asphalt binders. RMPD **15**(sup1), 78–112 (2014). https://doi.org/10.1080/14680629.2014.927411
3. Hugener, M., Kawakami, A.: Simulating repeated recycling of hot mix asphalt. RMPD **18**(sup2), 76–90 (2017). https://doi.org/10.1080/14680629.2017.1304263
4. Ekblad, J., Lundström, R.: Soft bitumen asphalt produced using RAP. Mater. Struct. **50**(1), 13 (2016). https://doi.org/10.1617/s11527-016-0871-z
5. EN 13108–8 (2016) Bituminous mixtures. Material specifications. Reclaimed asphalt. Brussels: European Committee for Standardization.
6. McDaniel R, Anderson R (2001) Recommended use of reclaimed asphalt pavement in the Superpave mix design method: technician's manual. Report 452, National Research Council, Transportation Research Board. Available online: http://onlinepubs.trb.org/onlinepubs/nchrp/nchrp_rpt_452.pdf Accessed 30 September 2020.
7. Circular Economy Action Plan (2020) European Commission, Brussels. Available online: https://ec.europa.eu/environment/circular-economy/pdf/new_circular_economy_action_plan.pdf Accessed 10 January 2021.
8. Poulikakos, L.D., Papadaskalopoulou, C., Hofko, B., Gschösser, F., Cannone Falchetto, A., Bueno, M., Arraigada, M., Sousa, J., Ruiz, R., Petit, C., Loizidou, M., Partl, M.N.: Harvesting the unexplored potential of European waste materials for road construction. Resour Conserv Recy **116**, 32–44 (2017). https://doi.org/10.1016/j.resconrec.2016.09.008
9. Partl MN, Bahia HU, Canestrari F, de la Roche C, Di Benedetto H, Piber H, Sybilski D (2013) Advances in interlaboraty testing and evaluation of bituminous materials. State-of-the-Art report of the RILEM Technical Committee 206-ATB. Springer, Dordrecht. https://doi.org/10.1007/978-94-007-5104-0
10. Partl MN, Porot L, Di Benedetto H, Canestrari F, Marsac P, Tebaldi G (2018) Testing and characterization of sustainable innovative bituminous materials and systems. State-of-the-Art Report of the RILEM Technical Committee 237-SIB. Springer International Publishing. https://doi.org/10.1007/978-3-319-71023-5
11. Poulikakos, L.D., Hofko, B., Cannone Falchetto, A., Porot, L., Ferrotti, G., Grenfell, J.: Recommendations of RILEM TC 252-CMB: relationship between laboratory short-term aging and performance of asphalt binder. Mater. Struct. **52**(4), 69 (2019). https://doi.org/10.1617/s11527-019-1370-9

12. Poulikakos, L.D., Cannone Falchetto, A., Wang, D., Porot, L., Hofko, B.: Impact of asphalt aging temperature on chemo-mechanics. RSC Adv. **9**(21), 11602–11613 (2019). https://doi.org/10.1039/C9RA00645A
13. Wang, D., Cannone Falchetto, A., Poulikakos, L., Hofko, B., Porot, L.: RILEM TC 252-CMB report: rheological modeling of asphalt binder under different short and long-term aging temperatures. Mater. Struct. **52**(4), 73 (2019). https://doi.org/10.1617/s11527-019-1371-8
14. Planche J-P, de la Roche C (2008) European survey on the use of RAP. In: Proceedings of ISAP conference, 18–20 August 2008, Zurich, Switzerland. Available online: Chantal de la Roche (rknet.it) Accessed 20 March 2021.
15. West RC, Copeland A (2015) High RAP asphalt pavements: Japan practice - lessons learned. National Asphalt Pavement Association (NAPA), Information Series 139. Available online: https://www.asphaltpavement.org/uploads/documents/EngineeringPubs/IS139_High_RAP_Asphalt_Pavements_Japan_Practice-lr.pdf Accessed on 14 February 2021.
16. Mohammadafzali, M., Ali, A.W., Sholar, G.A., Rilko, W.A., Baqersad, M.: Effects of rejuvenation and aging on binder homogeneity of recycled asphalt mixtures. J Transp Eng, Part B: Pavement **145**(1), 04018066 (2019). https://doi.org/10.1061/jpeodx.0000089
17. TL Asphalt-StB (2013) Technische Lieferbedingung für Asphaltmischgut für den Bau von Verkehrsflächenbefes-tigung. FGSV Verlag [in German]. Available online: https://www.stmb.bayern.de/assets/stmi/vum/strasse/bauunderhalt/49_tl_asphalt_2017.pdf Accessed 30 September 2020.
18. Petersen, J.C., Glaser, R.: Asphalt oxidation mechanisms and the role of oxidation products on age hardening revisited. RMPD **12**(4), 795–819 (2011). https://doi.org/10.1080/14680629.2011.9713895
19. Molenaar, A.A.A., Hagos, E.T., van de Ven, M.F.C.: Effects of aging on the mechanical characteristics of bituminous binders in PAC. J Mater Civil Eng **22**(8), 779–787 (2010). https://doi.org/10.1061/(ASCE)MT.1943-5533.0000021
20. Radenberg M, Boetcher S, Sedaghat N (2006) Effect and efficiency of rejuvenators on aged asphalt binder – German experiences. In: Proceedings of 6th Eurasphalt & Eurobitume Congress, 1–3 June 2016, Prague, Czech Republic.
21. Kawakami A, Kawashima Y, Nitta H, Yabu M (2019) An examination of property changes of repeatedly recycled asphalt bitumen using rejuvenator with high aromatic content. In, Cham. RILEM 252-CMB Symposium. Springer International Publishing, pp 189–194. https://doi.org/10.1007/978-3-030-00476-7_30
22. Porot L, Hugener M, Cannone Falchetto A, Wang D, Kawakami A, Hofko B, Grilli A, Pasquini E, Pasetto M, Tabatabaee H, Zhai H, Sa-da-Costa M, Soenen H, Kara De Maeijer P, Van den bergh W, Cardone F, Carter A, Vasconcelos K, Carbonneua X, Lorserie A, Mladenovic G, Oreskovic M, Koudelka T, Coufalik P, Bocci E, Zhang R, Dave E, Tebaldi G (2020) Aging of rejuvenated RAP binder - a RILEM inter-laboratory study. In: Proceedings of the 7th Eurasphalt & Eurobitume Congress v1.0 first published 1st July 2020 ISBN: 9789080288461.
23. Wang D, Cannone Falchetto A, Hugener M, Porot L, Kawakami A, Hofko B, Grilli A, Pasquini E, Pasetto M, Tabatabaee H, Zhai H, Sa-da-Costa M, Soenen H, Kara De Maeijer P, Van den bergh W, Cardone F, Carter A, Vasconcelos K, Carbonneua X, Lorserie A, Mladenovic G, Oreskovic M, Koudelka T, Coufalik P, Bocci E, Zhang R, Dave E, Tebaldi G (2020) Effect of aging on the rheological properties of blends of virgin and rejuvenated RA binders. In: Proceedings of RILEM International Symposium of Bituminous Materials, Lyon, France.
24. Karlsson, R., Isacsson, U.: Bitumen rejuvenator diffusion as influenced by ageing. RMPD **3**(2), 167–182 (2002). https://doi.org/10.1080/14680629.2002.9689920
25. Shen, J., Ohne, Y.: Determining rejuvenator content for recycling reclaimed asphalt pavement by SHRP binder specifications. Int. J. Pavement Eng. **3**(4), 261–268 (2002). https://doi.org/10.1080/1029843021000083685
26. Grilli, A., Gnisci, M.I., Bocci, M.: Effect of ageing process on bitumen and rejuvenated bitumen. Constr. Build. Mater. **136**, 474–481 (2017). https://doi.org/10.1016/j.conbuildmat.2017.01.027
27. Mogawer, W.S., Booshehrian, A., Vahidi, S., Austerman, A.J.: Evaluating the effect of rejuvenators on the degree of blending and performance of high RAP, RAS, and RAP/RAS mixtures. RMPD **14**(sup2), 193–213 (2013). https://doi.org/10.1080/14680629.2013.812836

28. Zaumanis, M., Mallick, R.B., Frank, R.: 100% recycled hot mix asphalt: A review and analysis. Resour Conserv Recy **92**, 230–245 (2014). https://doi.org/10.1016/j.resconrec.2014.07.007
29. Tabatabaee HA, Kurth TL (2016) Rejuvenation vs. softening: reversal of the impact of aging on asphalt thermo-rheological and damage resistance properties. In: Proceedings of ISAP Symposium, 18–21 July 2016 Jackson Hole, Wyoming US.
30. ASTM D2872–19 Standard test method for effect of heat and air on a moving film of asphalt (Rolling Thin-Film Oven Test). ASTM International, West Conshohocken, PA, USA. https://doi.org/ https://doi.org/10.1520/D2872-19
31. ASTM D6521–19a Standard practice for accelerated aging of asphalt binder using a pressurized aging vessel (PAV). ASTM International, West Conshohocken, PA, USA. . https://doi.org/10.1520/D6521-19A
32. EN 12607–1 (2014) Bitumen and bituminous binders - Determination of the resistance to hardening under influence of heat and air - Part 1: RTFOT method. Brussels: European Committee for Standardization.
33. EN 14769 (2012) Bitumen and bituminous binders - Accelerated long-term ageing conditioning by a Pressure Ageing Vessel (PAV). Brussels: European Committee for Standardization.
34. Porot L, Grady W (2016) Effectiveness of a bio-based additive to restore properties of aged asphalt binder. In: Proceedings of ISAP Symposium, 18–21 July 2016 Jackson Hole, Wyoming US.
35. Tabatabaee, H.A., Kurth, T.L.: Analytical investigation of the impact of a novel bio-based recycling agent on the colloidal stability of aged bitumen. RMPD **18**(sup2), 131–140 (2017). https://doi.org/10.1080/14680629.2017.1304257
36. BRRC (2020) Dossier 21 – Categorisation and analysis of rejuvenators for asphalt recycling. Available online: https://brrc.be/en/expertise/expertise-overview/dossier-21-categorisation-and-analysis-rejuvenators-asphalt-recycling Accessed 5 February 2021.
37. Nsengiyumva, G., Haghshenas, H.F., Kim, Y.-R., Kommidi, S.R.: Mechanical-chemical characterization of the effects of type, dosage, and treatment methods of rejuvenators in aged bituminous materials. Transport Res Rec **2674**(3), 126–138 (2020). https://doi.org/10.1177/0361198120909110
38. Epps A, Martin FW, Kaseer F, Arambula-Mercado E, Baja A, Garcia Cucalon L, Yin F, Chowdhury A, Epps J, Glover C, Hajj EY, Morian N, Sias Daniel J, Oshone M, Rahbar-Rastegar R, Ogbo C, King G (2020) Evaluating the effects of recycling agents on asphalt mixtures with high RAS and RAP binder ratios. NCHRP Report 927. Transportation Research Board, National Academies of Sciences, Engineering, Medicine. The National Academies Press, Washington, DC. https://doi.org/10.17226/25749
39. Haghshenas, D.F., Rea, R., Reinke, G., Haghshenas, D.F.: Chemical characterization of recycling agents. J Mater Civil Eng **32**(5), 06020005 (2020). https://doi.org/10.1061/(ASCE)MT.1943-5533.0003167
40. EAPA (2019) Recommendations for the use of rejuvenators in hot and warm asphalt production. Available online: https://www.baunetzwerk.biz/sites/default/files/2019-07/EAPA-Rejuvenators-Paper-web_0.pdf Accessed 30 September 2020.
41. Grilli, A., Bocci, E., Bocci, M.: Hot recycling of reclaimed asphalt using a bio-based additive. In: 8th RILEM International Symposium on Testing and Characterization of Sustainable and Innovative Bituminous Materials, pp. 953–964. Dordrecht. Springer, Netherlands (2016)
42. Nahar, S.N., Qiu, J., Schmets, A.J.M., Schlangen, E., Shirazi, M., van de Ven, M.F.C., Schitter, G., Scarpas, A.: Turning back time: rheological and microstructural assessment of rejuvenated bitumen. Transport Res Rec **2444**(1), 52–62 (2014). https://doi.org/10.3141/2444-06
43. Yu, X., Zaumanis, M., dos Santos, S., Poulikakos, L.D.: Rheological, microscopic, and chemical characterization of the rejuvenating effect on asphalt binders. Fuel **135**, 162–171 (2014). https://doi.org/10.1016/j.fuel.2014.06.038
44. Zaumanis, M., Mallick, R.B., Poulikakos, L., Frank, R.: Influence of six rejuvenators on the performance properties of Reclaimed Asphalt Pavement (RAP) binder and 100% recycled asphalt mixtures. Constr. Build. Mater. **71**, 538–550 (2014). https://doi.org/10.1016/j.conbuildmat.2014.08.073

45. Baghaee Moghaddam, T., Baaj, H.: The use of rejuvenating agents in production of recycled hot mix asphalt: A systematic review. Constr. Build. Mater. **114**, 805–816 (2016). https://doi.org/10.1016/j.conbuildmat.2016.04.015
46. Karki, P., Zhou, F.: Effect of rejuvenators on rheological, chemical, and aging properties of asphalt binders containing recycled binders. Transport Res Rec **2574**(1), 74–82 (2016). https://doi.org/10.3141/2574-08
47. Koudelka, T., Coufalik, P., Fiedler, J., Coufalikova, I., Varaus, M., Yin, F.: Rheological evaluation of asphalt blends at multiple rejuvenation and aging cycles. RMPD **20**(sup1), S3–S18 (2019). https://doi.org/10.1080/14680629.2019.1588150
48. Ali, A.W., Mehta, Y.A., Nolan, A., Purdy, C., Bennert, T.: Investigation of the impacts of aging and RAP percentages on effectiveness of asphalt binder rejuvenators. Constr. Build. Mater. **110**, 211–217 (2016). https://doi.org/10.1016/j.conbuildmat.2016.02.013
49. Bocci E, Prosperi E, P M (2020) Rheological modelling of the bitumen from reclaimed asphalt with rejuvenation and re-ageing. In: Proceedings of RILEM International Symposium of Bituminous Materials, Lyon, France.
50. Caputo P, Eskandarsefat S, Porto M, Ioise V, Oliveiro Rossi C, Scarpelli F, Venturini L (2020) Towards advances in differentiating recycling agents: softening agents vs. rejuvenating agents. In: Proceedings of the 7th Eurasphalt & Eurobitume Congress v1.0, first published 1st July 2020, ISBN: 9789080288461.
51. Li, Q., Li, G., Ma, X., Zhang, S.: Linear viscoelastic properties of warm-mix recycled asphalt binder, mastic, and fine aggregate matrix under different aging levels. Constr. Build. Mater. **192**, 99–109 (2018). https://doi.org/10.1016/j.conbuildmat.2018.10.085
52. Mazzoni, G., Bocci, E., Canestrari, F.: Influence of rejuvenators on bitumen ageing in hot recycled asphalt mixtures. J Traffic Transport Eng **5**(3), 157–168 (2018). https://doi.org/10.1016/j.jtte.2018.01.001
53. Ongel, A., Hugener, M.: Impact of rejuvenators on aging properties of bitumen. Constr. Build. Mater. **94**, 467–474 (2015). https://doi.org/10.1016/j.conbuildmat.2015.07.030
54. Yin, F., Kaseer, F., Arámbula-Mercado, E., Epps Martin, A.: Characterising the long-term rejuvenating effectiveness of recycling agents on asphalt blends and mixtures with high RAP and RAS contents. RMPD **18**(sup4), 273–292 (2017). https://doi.org/10.1080/14680629.2017.1389074
55. Zhou Z, Gu X, Dong Q, Ni F, Jiang Y (2020) Investigation of the oxidation ageing of RAP asphalt blend binders and mixtures. Int J Pavement Eng:1–17. https://doi.org/10.1080/10298436.2020.1763345
56. Behnood, A.: Application of rejuvenators to improve the rheological and mechanical properties of asphalt binders and mixtures: A review. J. Clean. Prod. **231**, 171–182 (2019). https://doi.org/10.1016/j.jclepro.2019.05.209
57. Xu, S., Liu, X., Tabaković, A., Lin, P., Zhang, Y., Nahar, S., Lommerts, B.J., Schlangen, E.: The role of rejuvenators in embedded damage healing for asphalt pavement. Mater Design **202**, 109564 (2021). https://doi.org/10.1016/j.matdes.2021.109564
58. ASTM D5, D5M–20,: Standard test method for penetration of bituminous materials. ASTM International, West Conshohocken, PA, USA (2020). https://doi.org/10.1520/D0005_D0005M-20
59. EN 1426 (2015) Bitumen and bituminous binders - Determination of needle penetration. Brussels: European Committee for Standardization.
60. EN 12697–3 (2013) European Standard. Bituminous mixtures. Test methods for hot mix asphalt. Bitumen re-covery: Rotary evaporator. Brussels: European Committee for Standardization.
61. AASHTO M320,: Standard specification for performance-graded asphalt binder. American Association of State Highway and Transportation Officials, USA (2017)
62. Chailleux E, Bessman E, Hornych P, Blanc J, Gaudefroy V, Sotoodeh-Nia Z, Manke N, Williams C, Cochran E, Lo Presti D, Jimenez A, Porot L, Planche J-P, Boysen RB, Pouget S, Olard F (2018) BioRePavation: innovation in bio-recycling of old asphalt pavements, comparison between EU and US mix design specification systems. In: 13th ISAP Conference on Asphalt Pavements, 19–22 June 2018 Foratleza, Ceara, Brazil. Paper 48.

63. Vansteenkiste S, Tanghe T, Duerinckx B, Vanelstraete A, Porot L, Govers B, Libbrecht D, Modde P-P (2020) Impact of bio-based rejuvenator on bitumen and asphalt mix performance - laboratory and field evaluation. In: Proceedings of the 7th Eurasphalt & Eurobitume Congress v1.0 first published 1st July 2020, ISBN: 9789080288461.
64. Cannone Falchetto A, Porot L, Riccardi C, Hugener M, Tebaldi G, Dave E Effects of rejuvenator on reclaimed asphalt binder: an exploratory study of the RILEM TC 264-RAP Task Group 3. In: Poulikakos LD, Cannone Falchetto A, Wistuba MP, Hofko B, Porot L, Di Benedetto H (eds) RILEM 252-CMB Symposium, 2019. Springer International Publishing, pp 195–200.
65. EN 1427 (2015) Bitumen and bituminous binders - Determination of the softening point - Ring and Ball method. Brussels: European Committee for Standardization.
66. ASTM D36, D36M–14,: Standard test method for softening point of bitumen (Ring-and-Ball Apparatus). ASTM International, West Conshohocken, PA, USA (2020). https://doi.org/10.1520/D0036_D0036M-14R20
67. EN 14770 (2012) Bitumen and bituminous binders - Determination of complex shear modulus and phase angle - Dynamic Shear Rheometer (DSR). Brussels: European Committee for Standardization.
68. AASHTO T315–20,: Standard method of test for determining the rheological properties of asphalt binder using a dynamic shear rheometer (DSR). American Association of State Highway and Transportation Officials, USA (2020)
69. ASTM D7175–15 Standard test method for determining the rheological properties of asphalt binder using a dynamic shear rheometer. ASTM International, West Conshohocken, PA, USA. https://doi.org/10.1520/D7175-15
70. EN 14771 (2012) Bitumen and bituminous binders - Determination of the flexural creep stiffness - Bending Beam Rheometer (BBR). Brussels: European Committee for Standardization.
71. ASTM D6648–08,: Standard test method for determining the flexural creep stiffness of asphalt binder using the Bending Beam Rheometer (BBR). ASTM International, West Conshohocken, PA, USA (2016). https://doi.org/10.1520/D6648-08R16
72. ASTM E168,: Standard practices for general techniques of infrared quantitative analysis (FTIR). ASTM International, West Conshohocken, PA, USA (2016). https://doi.org/10.1520/E0168-16
73. Farrar M, Sui C, Salmans S, Qin Q (2015) Determining the low-temperature rheological properties of asphalt binder using a dynamic shear rheometer (DSR). Report 4FP 08. Western Research Institute, USA.
74. Wang, D., Cannone Falchetto, A., Alisov, A., Schrader, J., Riccardi, C., Wistuba, M.P.: An alternative experimental method for measuring the low temperature rheological properties of asphalt binder by using 4mm parallel plates on dynamic shear rheometer. Transport Res Rec **2673**(3), 427–438 (2019). https://doi.org/10.1177/0361198119834912
75. Wang, D., Falchetto, A.C., Riccardi, C., Poulikakos, L., Hofko, B., Porot, L., Wistuba, M.P., Baaj, H., Mikhailenko, P., Moon, K.H.: Investigation on the combined effect of aging temperatures and cooling medium on rheological properties of asphalt binder based on DSR and BBR. RMPD **20**(sup1), S409–S433 (2019). https://doi.org/10.1080/14680629.2019.1589559
76. Wang D, Cannone Falchetto A, Riccardi C, Westerhoff J, Wistuba MP (2019) Investigation on the effect of physical hardening and aging temperature on low-temperature rheological properties of asphalt binder. RMPD:1–23. https://doi.org/10.1080/14680629.2019.1665089
77. Carret, J.-C., Cannone Falchetto, A., Marasteanu, M.O., Di Benedetto, H., Wistuba, M.P., Sauzeat, C.: Comparison of rheological parameters of asphalt binders obtained from bending beam rheometer and dynamic shear rheometer at low temperatures. RMPD **16**(sup1), 211–227 (2015). https://doi.org/10.1080/14680629.2015.1029696
78. Mouillet, V., Lamontagne, J., Durrieu, F., Planche, J.-P., Lapalu, L.: Infrared microscopy investigation of oxidation and phase evolution in bitumen modified with polymers. Fuel **87**(7), 1270–1280 (2008). https://doi.org/10.1016/j.fuel.2007.06.029
79. Lamontagne, J., Dumas, P., Mouillet, V., Kister, J.: Comparison by Fourier transform infrared (FTIR) spectroscopy of different ageing techniques: application to road bitumens. Fuel **80**(4), 483–488 (2001). https://doi.org/10.1016/S0016-2361(00)00121-6

Chapter 5
RAP Characterisation by Assessing the Relative Degree of Binder Activity

Davide Lo Presti, Kamilla Vasconcelos, Gustavo dos Santos Pinheiro, Ana Jiménez del Barco Carrión, Marko Orešković, Marco Pasetto, Emiliano Pasquini, Alan Carter, Alexandros Margaritis, Wim Van den Bergh, Xavier Carbonneau, Gurunath Guduru, Shanin Eskandarsefat, Francesco Preti, Eshan V. Dave, Gabriele Tebaldi, Gordon Airey, Nunzio Viscione, and Francesca Russo

D. L. Presti
University of Palermo, Palermo, Italy
e-mail: davide.lopresti@unipa.it

G. Airey
University of Nottingham, Nottingham, UK
e-mail: Gordon.airey@nottingham.ac.uk

K. Vasconcelos · G. dos Santos Pinheiro
Polytechnic School, University of São Paulo, São Paulo, Brazil
e-mail: kamilla.vasconcelos@usp.br

G. dos Santos Pinheiro
e-mail: gustavopnhr@usp.br

A. J. del Barco Carrión
University of Granada, Granada, Spain
e-mail: gaspare.giancontieri@you.unipa.it

M. Orešković
Faculty of Civil Engineering, University of Belgrade, Belgrade, Serbia

M. Pasetto · E. Pasquini
University of Padua, Padua, Italy
e-mail: marco.pasetto@unipd.it

E. Pasquini
e-mail: emiliano.pasquini@unipd.it

A. Carter
École de Technologie Supérieure, Montreal, Canada
e-mail: Alan.Carter@etsmtl.ca

G. Tebaldi and E. Dave (eds.), *Recycling Asphalt Pavement*, RILEM State-of-the-Art Reports 44, https://doi.org/10.1007/978-3-031-94231-0_5

5.1 Motivation

Asphalt recycling has been proven to be one of the best practices for improving the sustainability of asphalt pavements. It has several advantages, including saving natural resources of aggregate and bitumen, reducing environmental impacts resulting from the production of virgin raw materials, decreasing the initial cost of asphalt, and saving the environment from solid wastes. Despite these benefits, many challenges still hinder the implementation of asphalt mixes with high reclaimed pavement content (RAP) content. One of the main challenges is the understanding and characterisation of the interaction between the RAP aged binder and the necessary recycling agents (e.g. soft binders, rejuvenators).

Design of asphalt mixtures with RAP requires accurate characterisation of RAP properties such as bitumen content, bitumen properties, aggregate properties, and gradation. It is a conventional procedure to assume that all the RAP binder will be available in the final mix, hence, to assume a full blending between the RAP binder and the recycling agent. However, it is widely accepted that the full blending

A. Margaritis · W. Van den Bergh
Universiteit Antwerpen, Antwerpen, Belgium
e-mail: alexandros.margaritis@uantwerpen.be

W. Van den Bergh
e-mail: wim.vandenbergh@uantwerpen.be

X. Carbonneau
Colas, Paris, France
e-mail: carbonneau@campus.colas.fr

G. Guduru
Indian Institute of Technology Kharagpur, Kharagpur, India

S. Eskandarsefat
Iterchimica S.R.L., Suisio, Italy
e-mail: shahin.eskandarsefat@iterchimica.it

F. Preti
University of Parma, Parma, Italy
e-mail: francesco.preti@studenti.unipr.it

F. Preti · E. V. Dave (✉) · G. Tebaldi (✉)
University of New Hampshire, Durham, USA
e-mail: eshan.dave@unh.edu

G. Tebaldi
e-mail: gtebaldi@unipr.it

N. Viscione · F. Russo
Department of Civil Environmental and Construction Engineering, University of Naples Federico II, Naples, Italy
e-mail: nunzio.viscione@unina.it

F. Russo
e-mail: francesca.russo2@unina.it

assumption is rarely true. In fact, some of the RAP binder, especially if heavily aged, might not function as a bounding material in a new mix. Consequently, asphalt technologists are still unable to quantify or estimate the amount of RAP binder they can consider as a replacement of virgin binder added in the new mix design process.

The motivation of the task group is to develop a simple and reliable method to characterise RAP by estimating the so-called Degree of binder Activity (DoA) via testing 100% RAP mixes without using any recycling agents. Asphalt technologists can use DoA to improve mix design procedures by estimating the minimum amount of RAP binder that could be considered available for hot, warm and half-warm asphalt mixes.

5.2 Background

5.2.1 Introduction

Asphalt recycling dates back to 1915; however, in 1970, it started receiving more attention due to the inflation of oil prices caused by the Arab oil ban [1]. The aim was to recycle existing asphalt to reduce the construction cost of asphalt pavements. RAP can be considered a valuable alternative source of bitumen and aggregate because recycling this material in asphalt mixtures reduces the use of virgin bitumen and aggregate [2]. RAP has several valuable economic and environmental benefits, including [1, 3]:

1. Reducing the production cost of asphalt mixes: In the Hot mix asphalt industry, material cost comprises 70% of the total production cost of asphalt [2]. However, when RAP is used, a significant reduction in the material cost is recorded, as shown in Fig. 5.1 [4]. This is mainly due to the reuse of aged bitumen as a replacement of the virgin bitumen in the new asphalt mixtures.
2. Increaseing the resource efficiency while reducing the consumption of virgin aggregates and bitumen: When RAP is recycled, RAP binder and aggregate reduce the amount of virgin binder and aggregate required for the new asphalt mix.
3. Maintaining the in-service road geometrics: When asphalt is milled and recycled, it allows for laying a new pavement layer without significantly altering the pavement profile. This is particularly important when maintaining existing pavement geometry is required, such as for clearance under bridges.
4. Protecting the environment: Aged asphalt is solid waste which may end up in landfills unless it is recycled using the best available practices. Moreover, recycling RAP means significantly reducing the negative environmental impact resulting from producing raw bitumen and aggregate. This is because, during the production phase of the raw materials, a massive amount of pollutants is emitted into the air due to burning fuel required to operate machines and equipment used to produce these materials. Lee et al. [5] stated that a reduction of 20% in

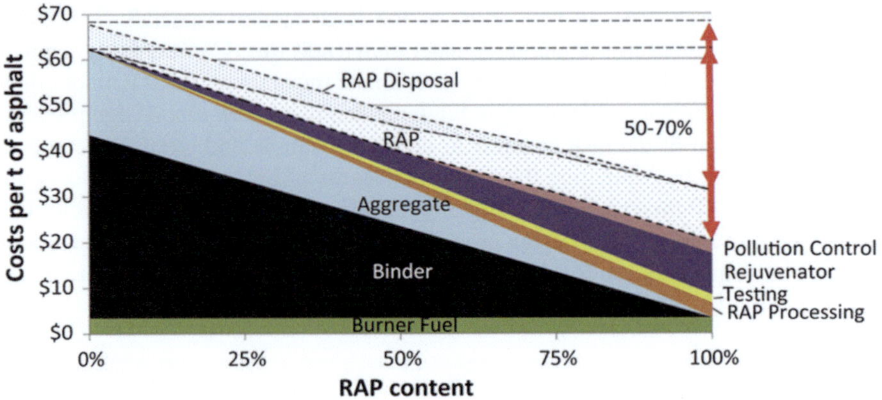

Fig. 5.1 Asphalt costs and RAP economic benefits [4]

CO_2 emissions is achieved when using recycled materials as a result of reduced emissions resulting from the production phase of these materials In contrast, Zaumanis et al. [6] stated that CO_2 reduction could be increased up to 35% when 100% RAP is recycled.

5. Reduction of Energy Consumption: Using RAP would help reduce the amount of virgin materials required in new asphalt mixtures, which reduces the energy required to produce the raw materials. This means that considerable quantities of fuel required for materials production will be saved, which constitutes a significant positive economic impact on the asphalt industry. Lee et al. [5] stated that 16% less energy is required when RAP is recycled.

RAP recycling benefits have multiple economic and environmental benefits. However, there are also some factors and technical issues that hinder or reduce the incorporation of this material in new asphalt mixtures, including [4, 7, 8]:

1. Variability of RAP properties among different sources and even within the same source of RAP.
2. Uncertainty regarding the performance of recycled asphalt mixtures, especially those containing high RAP content.
3. The increased complexity of the mix design when RAP is incorporated.
4. Technical issues regarding RAP incorporation and the capacity of asphalt plants often limit RAP recycling to a maximum of about 50%.

The mistrust in the performance of asphalt mixtures containing RAP is a critical point stressed in the literature and has led to limiting RAP incorporation levels. On fatigue cracking performance, several researchers have pointed out that RAP incorporation makes asphalt mixtures susceptible to that kind of failure [9, 10]. This drop in fatigue performance has been attributed to the ageing and stiffness of the RAP binder. It is believed that the high stiffness of the RAP binder makes it lose its elastic recovery properties, which eventually reflects on fatigue resistance performance. On low-temperature cracking performance, the presence of RAP binder changes the failure mode from ductile to brittle; it also increases the critical low-temperature cracking [11, 12]. This means that asphalt pavements containing RAP are more prone to exhibit low temperature cracking at higher temperatures than asphalt mixtures composed of virgin materials.

5.2.2 *RAP Characterisation Methods*

RAP can be incorporated into new asphalt mixtures to decrease the amount of virgin binder and aggregate. Before incorporating RAP, however, it is essential to identify its engineering properties and include them in the mix design process to assure a comparable performance to conventional HMA. These include, among others, RAP binder content, RAP aggregate particle size distribution, RAP aggregate consensus properties such as the angularity of the coarse and fine particles, and RAP binder properties [13]:

- The amount of available RAP binder is a direct input to the design of recycled asphalt mixes; in fact, it is fundamental for the development of blending charts and governs the amount of recycling agents which should be added in mix design exercises. In order to establish the level of participation of the RAP binder, it is paramount to determine the RAP binder content. For instance, a minimum of 3.8% RAP binder content is required in Japan when RAP is incorporated into new asphalt mixtures [14]. This is probably because, below that level, the participation of RAP binder in the total mix will be lower than expected, which might result in dry asphalt mixes.
- RAP aggregate gradation is also a direct input in the asphalt mix design process. The particle size distribution of RAP aggregate is required to comply with a particular standard aggregate gradation. Therefore, when combining RAP aggregate with virgin aggregate, the overall grading meets the grading requirements of the mix design standard.
- The properties of RAP aggregate particles, such as crushed aggregate content, aggregate angularity, and dust proportion, are essential indicators of the effect of RAP aggregate on critical measures such as mix volumetric properties and performance [11]. However, these properties should be evaluated after blending RAP aggregate with virgin aggregate to meet the required properties of combined aggregate rather than RAP aggregate alone [15].
- Another critical property of RAP materials is their aged binder's characteristics. Bitumen ageing is a complex physicochemical process that takes place over time, starting from the stage of heating and mixing asphalt at a plant until the end of asphalt life. Researchers have identified different mechanisms for bitumen ageing, such as oxidation, evaporation, and absorption [16, 17]. The overall effect of bitumen ageing is an increase in binder stiffness and loss of bitumen recovery properties, making asphalt susceptible to particular distress types such as fatigue cracking and low temperature cracking [7, 18]. This means RAP binder must

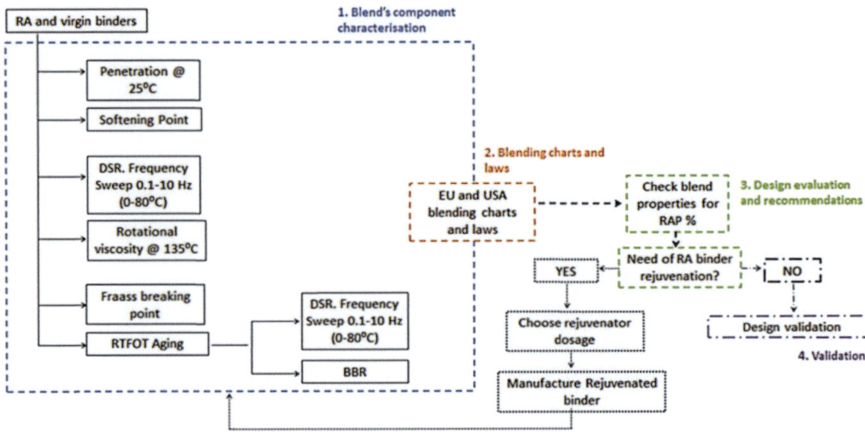

Fig. 5.2 RAP binder characterisation methodology [8]

be rejuvenated to restore its initial properties in order to achieve comparable performance to conventional HMA. One of the widely applied methods is the use of a softer bitumen as a recycling agent. When the binders blend, the softer binder rejuvenates the RAP binder, whereas the aged binder stiffens the soft binder [11]. The blending ratio between the two binders strongly depends on the properties of the RAP binder. Accordingly, it is essential to characterise the RAP binder before commencing any mix design. Other RAP binder characteristics include bitumen penetration grade, softening point, rheological properties, and low-temperature behaviour, as shown in Fig. 5.2. This methodology was suggested by Lo Presti et al. [8] with the aim to fully characterise RAP bitumen and the design binder blend in the mix. Based on the RAP incorporation level, all or some of these properties are required to design the bitumen blend of the RAP and virgin binders.

5.2.3 RAP Incorporation Level and Mix Design

RAP incorporation level differs from country to country based on the legislation and local experience with RAP recycling methods. It also depends on the structural layer that RAP is recycled into. In surface layers, recycling levels are still low at around 15% because these layers are in direct contact with traffic loading and environmental conditions; therefore, they require high distress resistance [19]. More RAP can be incorporated in the lower layers, up to 50% [1]. Nevertheless, many countries have generally limited RAP recycling in asphaltic mixtures to 10–30%. Until 2006, RAP incorporation in asphalt mixtures was limited to approximately 15% because at this level of RAP, no binder grade change was required, which reduces the costs of binder testing to characterise RAP binder properties [1]. However, due to the economic and environmental benefits of RAP recycling and the increasing costs of virgin bitumen and aggregate, highway agencies have become more willing to increase the level of RAP incorporation. In 2014, the average RAP recycling level in the US was about 20% [20]. More recently, different studies have demonstrated the feasibility and capability of incorporating 100% RAP in asphalt mixtures [4, 8, 21].

From a binder blending point of view, the level of incorporation is directly related to the percentage of RAP, which determines the percentage of the stiff binder in the mix. The presence of aged bitumen can change the mechanical properties of the designed mix due to its effect on the properties of the soft binder when the two binders are blended. Accordingly, highway agencies have set some guidelines to adjust mix design in case of RAP incorporation. According to the Superpave mix design specifications for RAP incorporation, three levels of RAP were determined: less than 15%, 15–25%, and more than 25% [11]. The low RAP dosage requires no grade change of virgin binder as the RAP binder content is relatively low in this case and will minorly affect the properties of the virgin binder. The intermediate RAP incorporation level involves reducing the virgin binder grade by one increment (6 °C based on the PG grading system) at both the high and low-performance grades. However, at higher RAP dosages, blending charts are recommended as the effect of RAP binder becomes significant due to its increased percentage in the mix. Later, the Superpave specifications were slightly modified, and the RAP binder's low PG was considered when designing virgin binder grade [15], as shown in Table 5.1. The critical low temperature of the RAP binder was found to be an important factor in limiting RAP percentage. The higher the low critical temperature, the stiffer the RAP binder, and the higher the susceptibility of the designed mix to low temperature cracking. However, Al-Qadi et al. [22] conducted a study to determine the amount of active RAP binder when different percentages of RAP (0, 20, and 40%) are recycled; they concluded that incorporating up to 20% of RAP did not require adding a softer bitumen, and up to 40% of RAP required two PG levels softer bitumen. More recently, McDaniel et al. [23] showed that recycling up to 25% RAP did not require a change in the virgin binder grade. They also stated that this recycling method did not adversely affect the low-temperature performance of that mix. In Europe, EN standard 13,108-1-2006 determines the level of RAP recycling based on the structural layer that will be constructed. For surface courses, 10% of RAP is allowed without any design modification, whereas 20% is allowed in binder and base courses.

Table 5.1 Superpave specifications for RAP mix design

Recommended virgin binder grade	Recovered RAP binder low critical temperature °C		
	−22	−16	−10
	RAP %		
No binder grade change	< 20	< 15	< 10
One PG grade softer	20–30	15–25	10–15
Use blending charts	> 30	> 25	> 15

From a mix design point of view, there are two ways to design asphalt mixtures containing RAP. The first is to design based on a known RAP percentage; the second is to design based on a known virgin binder grade [24]. The first method allows the designer to select the RAP incorporation level based on the desired criteria,then, based on that, binder blending can be designed, and the soft binder grade can be determined. The second method is useful when the soft binder properties are known because they can be used to calculate the percentage of RAP incorporated in the mix. Both methods require the construction of binder blending charts when the RAP percentage in the mix exceeds threshold limits identified by highway agencies. To construct binder blending charts, however, there are two frequently used methods, as follows:

5.2.3.1 Superpave Method

The NCHRP report 452 [15] describes two procedures to design asphalt mixes containing RAP, one for a predetermined RAP content and one for identified virgin binder properties. The RAP binder should be extracted and recovered for the known RAP recycling level, and its critical high and low temperatures should be determined. Then the Performance Grade (PG) of the virgin binder can be calculated using the following equation:

$$T_{virgin} = \frac{T_{blend} - (RAP\% \times T_{RAP})}{(1 - RAP\%)} \tag{5.1}$$

where T_{virgin} is the virgin binder critical temperature that should be calculated at high and low temperatures to determine the required virgin binder PG, RAP% is the content of RAP in the mix, T_{blend} is the critical temperature of the final binders blend and T_{RAP} is the corresponding critical temperature of the RAP binder. For the known virgin binder properties, the RAP incorporation level that satisfies the PG requirements of the target binder grade can be determined as follows:

$$RAP\% = \frac{T_{blend} - T_{virgin}}{T_{RAP} - T_{virgin}} \tag{5.2}$$

5.2.3.2 European Method

In Europe, usually, if more than 10% RAP is recycled in surface courses or more than 20% in binder and base courses, then the EN standard 13,108-1 states that either the penetration or the softening point of the resulting binder should satisfy the requirement of the selected paving grade bitumen, as follows:

$$\log pen_{blend} = a * \log pen_{RAP\ binder} + b * \log pen_{virgin\ binder} \tag{5.3}$$

$$T_{R\&Mmix} = a \times T_{R\&B\,RAP\,binder} + b \times T_{R\&B\,virgin\,binder} \quad (5.4)$$

where a and b are the proportions by mass of the RAP binder and the virgin binder with respect to the total binder content (a + b = 1), pen_X is the penetration grade of designated binder, and $T_{R\&M\,mix}$ is the softening point of the designated binder.

5.2.4 RAP Microanalysis

Understanding the interaction between RAP and virgin materials is an important step in mixing asphalt mixtures with RAP. RAP binder properties and behaviour play an important role in the performance of the final mix. When mixing RAP with virgin materials, the RAP is conditioned at a certain temperature for a certain time to dry and to reduce RAP binder viscosity. During mixing, RAP particles are mechanically mixed with the virgin aggregate and bitumen, and the RAP binder becomes softer due to the diffusion of the soft binder. This diffusion process depends on several critical factors, including RAP content in the mix, RAP binder grade, RAP binder content, virgin binder grade, mixing time, mixing temperature, and RAP conditioning temperature. The level of diffusion or blending between the two binders controls to a high extent, the validity of some assumptions concerning the behaviour of the RAP in the mix. If all of the RAP binder will be blended, then the black rock assumption is invalid. However, if no blending occurs between the two binders, which is an impossible assumption, then the RAP can be assumed as black rock. Realistically, RAP behaviour is between these two extremes, and a large percentage of the RAP binder will probably be active in the mix (Fig. 5.3). However, there is a need to answer some questions concerning the activation and blending of the RAP binder. What percentage of RAP binder will be active at the selected production conditions? What is the degree of blending between the active RAP binder and the soft binder? The two following sub-sections provide a deeper insight into these concepts.

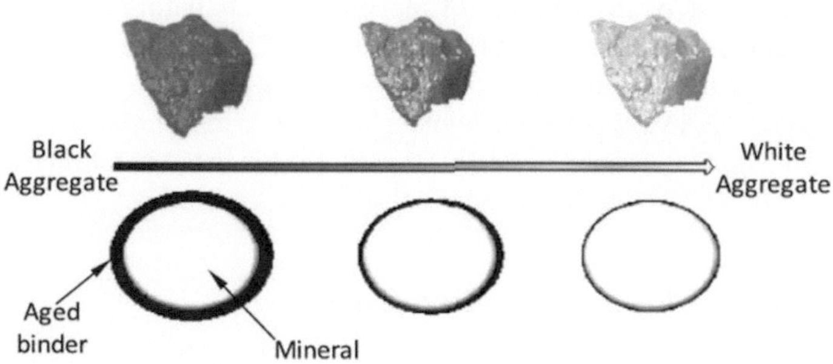

Fig. 5.3 Black and white rock concepts [25]

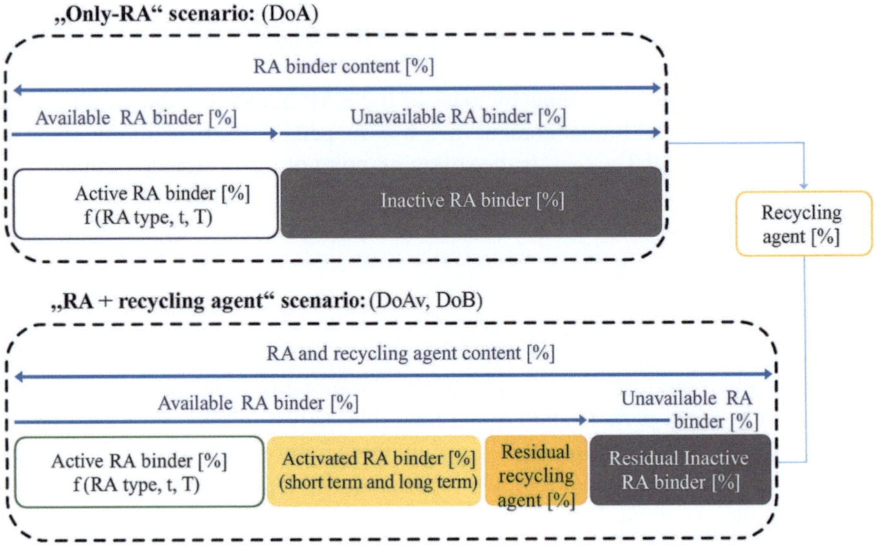

Fig. 5.4 Schemicatic representation of the DoA and DoB concepts [25]

5.2.5 *Degree of Binder Activity (DoA)*

According to the theoretical framework in Fig. 5.4 [25], the RAP binder surrounding an aggregate can be divided into two types: the first one is an outer layer of aged binder that can be activated, functioning as a bounding material, whenever RAP is exposed to hot mixing temperatures for some time, the second one, the inactive binder, it is represented by a series of aged binder layers that can eventually be "activated" only with the use of recycling agents (along with heat). On this basis, the DoA, defined as the percentage of the active binder weight to the total weight of the binder in the RAP, can be considered an intrinsic property of RAP. It will vary depending on the RAP type and processing conditions (i.e. mixing temperature and time), regardless of the addition of recycling agents before and/or during the Reclaimed Asphalt mixtures (RAM) manufacturing [25]. Hence, DoA could label RAP by providing the minimum amount of available aged binder to be considered in designing the new RAM. A complete nomenclature of terms related to the theoretical framework of Fig. 5.4, defined in Lo Presti et al. [25] is listed in Glossary.

5.2.6 *Degree of Binder Blending (DoB)*

In the scenario of blending a recycling agent with RAP binder, the interaction between the two binders is a critical factor that significantly impacts the mix design and mechanical properties. If the two binders reach a full blending during an asphalt mixing process, then the two binders perform comparably to the target binder resulting from the blending process. However, if no blending occurs during mixing, the resulting binder will be inhomogeneous and mix performance will be affected by the properties of the two binders rather than one thoroughly blended binder.

Theoretically, this phenomenon has been described in the literature as the Degree of Blending (DoB); it is basically a term used to describe the quality of the blending between the RAP and virgin binders. The DoB is entirely related to how RAP binder acts in the mix, whether it is a very aged binder, an active binder that cannot be blended, or an active binder that can be blended with the rejuvenator. In fact, the RAP incorporation limits discussed in the previous section are based on the expected DoB at different RAP percentages. The DoB is expected to be insignificant at low RAP recycling levels because there is a small amount of RAP binder to affect the properties of the virgin binder [11]. Therefore, no binder change is required. However, limited DoB can be achieved during mixing at intermediate RAP dosages, so one grade softer binder is recommended to offset the effect of binder blending on the resulting properties. However, at high RAP recycling levels, highway agencies recommend using blending charts to determine the grade of the virgin binder [26]. This suggests that the DoB is about 100% because the virgin binder fully interacts with the hard binder, and the resulting binder controls the mechanical performance of the mix.

Regardless of RAP percentage, DoB is theoretically between 0–100%, where 0% stands for the black rock case, and 100% represents the full blending scenario [27]. Figure 5.5 graphically illustrates the DoB at different levels. However, the black rock assumption has been disproved by several studies [11, 15, 24, 28] as these studies have clearly demonstrated that there is a blending level between the RAP binder and the virgin binder. Accordingly, researchers have investigated this critical topic and suggested different methods to quantify the DoB and its consequences on the mix design process, as Orešković et al. [29] summarized.

Jiménez del Barco Carrión et al. [18] and Lo Presti et al. [8] suggested including the DoB at a theoretical level assessing its influence in binders' properties in asphalt mixtures with high RAP content to determine what DoB value would be acceptable to obtain the desired properties of the blend.

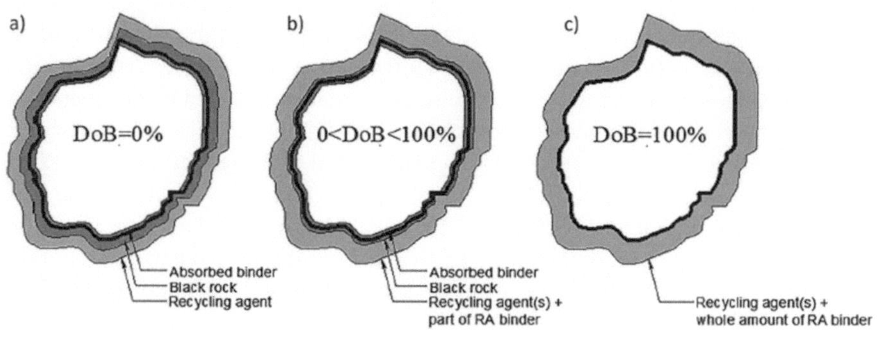

Fig. 5.5 DoB graphical explanation [25]

Bonaquist [30] developed a method to assess DoB by dynamic modulus measurements. The method stated that the RAP binder should be extracted, recovered, and blended with the virgin binder. The resulting binder should be characterised, and the mix dynamic modulus is estimated using the Hirsh model based on binder characteristics and mix volumetric properties. After that, the actual dynamic modulus of the mix is determined by running dynamic modulus testing; if there is sufficient overlap between the estimated and measured dynamic modulus master curves, then the DoB is sufficient. Otherwise, there is no full blending between the two binders. This method, however, requires RAP binder extraction, recovery, characterisation, and asphalt dynamic modulus testing, which is not available in all pavement materials laboratories. Also, the dynamic modulus results may not reflect DoB in the mix due to the compressive forces among aggregate particles [22], which means the aggregate may affect the test results and, eventually, the DoB results.

Another method was developed by Shirodkar et al. [27]. They suggested an iterative procedure to calculate the DoB based on the Superpave critical high-temperature parameter G*/sinδ for binders recovered from predefined sizes of RAP and virgin aggregate particles. Their method was to assume an initial percentage of RAP binder that will be active in the mix, design and manufacture samples based on this assumption, and then sieve and isolate RAP from aggregate particles and extract and recover binder from both portions to run DSR testing. The DoB was then determined using the following:

$$DoB(\%) = 100 \times \left| 1 - \frac{|CHTP_{vab} - CHTP_{rab}|}{|CHTP_{vb} - CHTP_{rb}|} \right| \tag{5.5}$$

CHTP is the critical high-temperature parameter (G*/sinδ) for the designated binder, *vab* is the binder recovered from the virgin aggregate, *rab* is the binder recovered from RAP aggregate, *vb* is the virgin binder, and *rb* is the RAP binder. If the calculated DoB is close to the assumed percentage of RAP binder participation, then the correct DoB has been obtained; if not, then a new value for RAP binder participation has to be assumed, and DoB has to be calculated again until these two values converge. They reported 70% and 96% DoB for 25% and 35% RAP incorporation, respectively. The limitation of this method is that at least three iterations are required before obtaining a reliable DoB result. Also, this procedure is to determine DoB at one mixing condition. Since DoB is a function of different production conditions such as mixing time, temperature, rejuvenator, and aged binder properties [31], this procedure is impractical as it takes a lot of trials and material to study the effects of production conditions on DoB. Coffey et al. [32] defined the DoB as the mobilisation percentage of RAP binder in an asphalt mix. They estimated the DoB based on air voids content by assuming the initial DoB is 70% based on literature; if this assumption was valid, then the air voids at 75 gyrations have to be 4%; otherwise, the binder content must be adjusted based on a revised DoB percentage. They reported a DoB of 84.7–90% for three different RAP sources with 25% RAP incorporation. However, this level of RAP incorporation was relatively low and reported in the literature to have a minor impact on the design mix. Also, this method requires some iterations until the assumed and calculated DoB converge.

Furthermore, the DoB is not a constant property; different factors can affect and change it. On the one hand, the mobilisation of RAP binder during mixing is significantly related to its viscosity. Since the viscosity depends on the binder grade and temperature, logically, mixing temperature should significantly impact its viscosity DoB. On the other hand, mixing time is another critical factor affecting the DoB by mechanical blending during asphalt mixing. Increasing mixing time is expected to increase the blending as there is more time for the binders to contact and interact with each other while their viscosities are still relatively low.

Navaro et al. [33] studied the effect of the mixing time and temperature on binder blending of an asphalt mixture containing 70% RAP produced at three temperatures, 110, 130, and 160 °C and mixed for different times between 20 s and 10 min. They studied binder blending by an image processing method. They reported that these two parameters have a significant impact on the homogeneity of the mix. They demonstrated that decreasing the mixing temperature from 160 to 130 °C required doubling to tripling mixing time to produce a mix with the same homogeneity. Bowers et al. [34] studied the effect of the same parameters on the blending and diffusion process between RAP and virgin binders. Predefined RAP and virgin aggregate sizes (RAP aggregate between 9.51–2.38 and virgin aggregate between 19–12.7 mm) were used in mixing. Binders from these aggregate particles were extracted through staged extraction to characterise the properties of the binders extracted from different layers. Gel Permeation Chromatography was used to study the recovered binders' properties based on the bitumen's molecular weight distribution. They concluded that mixing temperature has a critical effect on DoB, whereas increasing the mixing time up to 2.5 min improved the DoB, but beyond that, there was a limited impact on the DoB. The effect of mixing time and temperature on the DoB was also qualitatively evaluated in a different study by Margaritis et al. [35]. The bituminous coating of RAP particles was directly evaluated using an FTIR microscope. It was concluded that higher temperature and longer mixing time (5 min and 180 °C) drastically improved the blending efficiency compared to the sample prepared using shorter mixing time and lower temperature (2 min and 150 °C). Hassan et al. [36] studied the effect of mixing time on the homogeneity and particularly the distribution of air voids in asphalt samples containing 15% RAP. X-ray Computed Tomography and optical microscopy were used in this study to quantify and analyse air voids. The researchers stated that increasing up to double the mixing time resulted in a better air voids distribution in the studied specimens.

5.3 Literature Review Summary

To sum up, the DoB is a critical factor that affects the volumetric properties and the mechanical performance of mixtures containing RAP. It is a complex function of several factors, including RAP%, RAP binder properties, and production conditions of the asphalt. It must be highlighted here that DoB and DoA are indirectly related. The DoB can be considered an indicator of the blending quality of the active RAP binder and the recycling agent. In contrast, the DoA is an intrinsic property of RAP and can be used to estimate the minimum amount of the active RAP binder. DoA is assessed for each RAP type in a scenario with no recycling agent used. If, instead, a recycling agent (softer binder, a lubricant or a rejuvenator) is used, then the DoA will increase due to an enhanced activation level. Furthermore, it can be realized that much attention has been paid in the literature to the interaction of RAP binder with virgin recycling agent, however, limited attention has been paid to characterise RAP based on the potential to provide aged bitumen in a new mix formula. This topic needs then further research, and it is the focus of the interlaboratory testing exercise of the TG5.

5.4 Materials and Methods

5.4.1 RAP Sources and Characteristics

As part of this interlaboratory effort, 17 labs participated, named LAB1 to LAB17, with a total of 42 RAPs analyzed from 10 countries (Belgium, Brazil, Canada, France, India, Italy, Serbia, Spain, UK, and USA). They vary in terms of binder content/characteristics and gradation curves (black and white curves). Figure 5.6 shows all the gradations. The difference between the black and white curve for some RAP sources is shown in Fig. 5.7.

5.4.2 Interlaboratory Approach

5.4.2.1 RAP Selection and Preliminary Analysis

Suitable quantities of virgin materials and RAP were collected (Table 5.2). The RAP was mainly derived from the following:

- Three types of asphalt mixtures: Dense grade (Asphalt Concrete, AC), Gap graded (Stone Matrix Asphalt, SMA), and Open-graded (Porous Asphalt, PA).
- Two sub-types of RAP will be found containing neat bitumen and PMB, and;
- Two sub-types based on the different stages of ageing level as short-term aged (STA, usually with pen > 20) and long-term aged (LTA, usually with pen < 10).

The first operation was focused on the RAP gradation and binder content. These determinations were developed according to the composition analysis procedure carried out on each RAP. The maximum density is also determined. Not less important, the percentage of fines present in the material was also obtained to be used as a

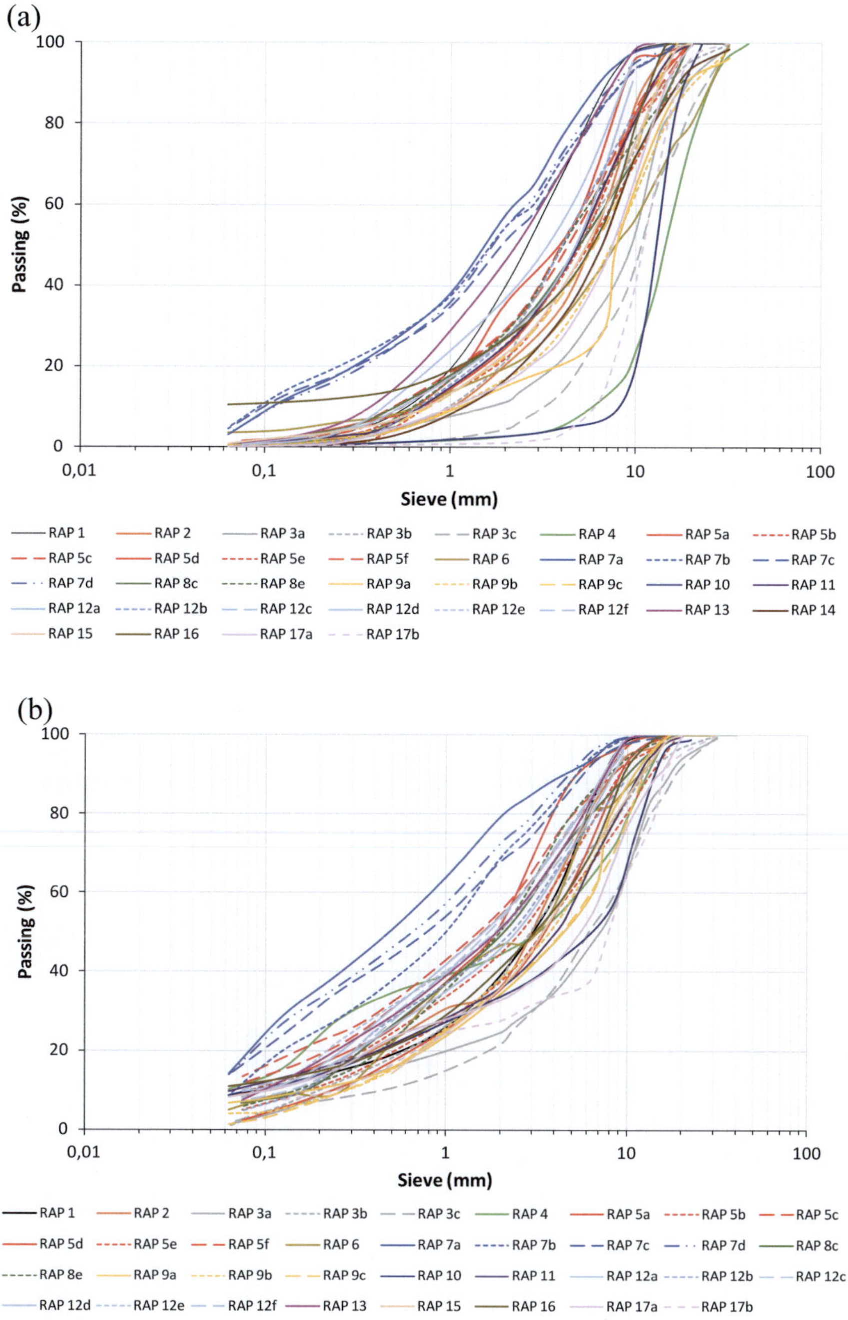

Fig. 5.6 RAP **a** black curve, and **b** white curve

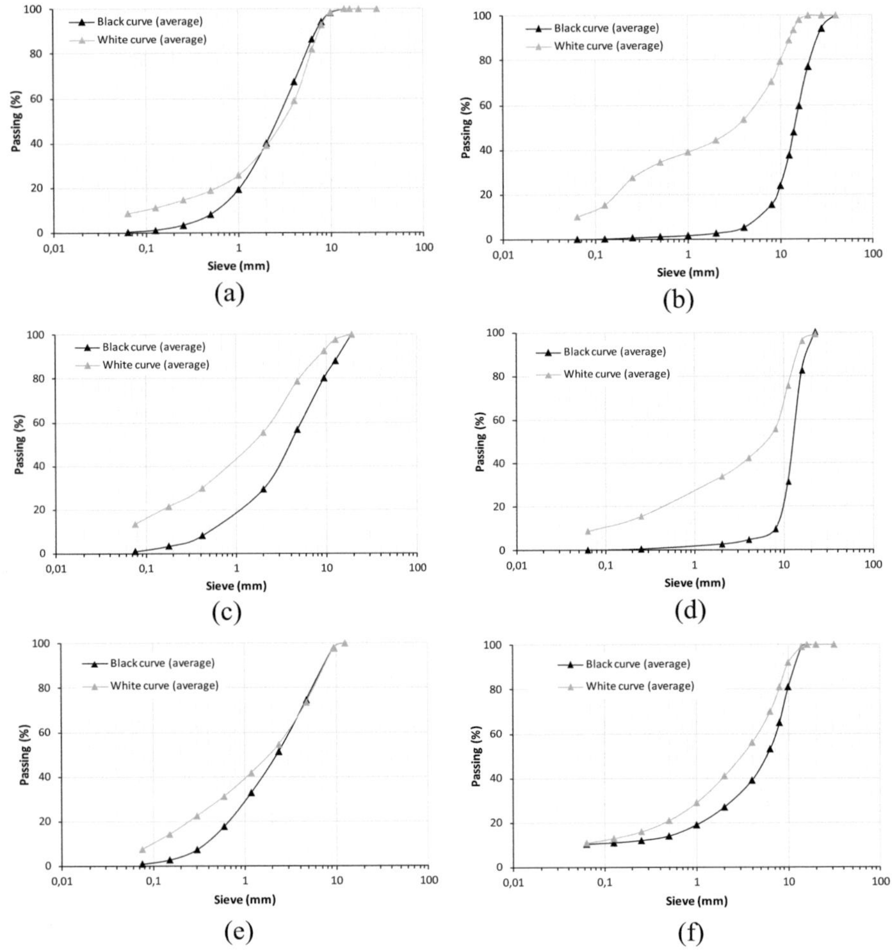

Fig. 5.7 **a** RAP 1, **b** RAP 4, **c** RAP 5c, **d** RAP 10, **e** RAP 13, **f** RAP 16

Table 5.2 RAP types

	STA—Virgin	STA—PMB	LTA—neat	LTA—PMB
AC	X	X	X	X
SMA	X	X	X	X
PA	X	X	X	X

threshold to use the reference materials. The selected RAP materials were subjected to a binder recovery procedure detailed in the standard EN 12,697-4 Bitumen recovery: Fractionating column; the bitumen extracted was then tested in a range of characterisation tests, such as Needle Penetration, Softening Point, Viscosity, etc., in order to identify key properties to be applied in the main process of the present task. The results obtained from those recovered binders could be correlated with the outcomes that originated in the RAP classification according to mechanical properties.

5.4.2.2 TG5 (DoA') Methodology

In these tasks, the selected RAPs will be subjected to a procedure developed at the University of Nottingham (UNOTT) for this TG5. This procedure allows for finding the amount of RAP binder that can be considered active in a new mixture formulation [37]. This is currently identified as the Relative Degree of Binder Activity (DoA'). The DoA' is very variable, but that has been shown to vary significantly with temperature and RAP type. Within AllBack2Pave [38], it was suggested to assume the DoA between 60 and 100%. For this interlaboratory tests exercise, TG5 would like to provide a methodology that allows providing an estimate of the relative DoA (DoA') by linking it to commonly used mechanical parameters of asphalt mixes, as presented in Fig. 5.8. Further discussion on the DoA' terminology is presented in item 5.1.

- The RAP is firstly dried in a ventilated oven at 40 °C for 48 h;
- Then the material is properly homogenized using a rifle box to obtain a sample;
- Pre-condition the RAP sample in the oven for 4 h (prior to mixing) at the compaction temperatures;
- Mechanical mixing/or hand mixing of the samples is conducted while controlling the temperature;
- The specimens are compacted using a Marshall compactor, with 50 blows on each side.

This procedure was adopted based on the findings of previous studies [39]. The Marshall moulds were used for compaction. The usual sizes of its specimens are 100 mm in diameter and 63.5 mm in height, which is the initially targeted height in Marshall Mix Design Method. As the maximum density of the material was determined, and in order to reach the height, the air voids content is estimated according to previous results from TG6 of RILEM TC SIB [40], then it is possible to determine the amount of material that should be placed in each mould for further compaction.

The RAP labelling through the DoA' procedure is then based on testing 100% RAP samples with selected mechanical tests available at each laboratory. The specimen manufacturing procedure was identical for each partner. However, both RAP types and mechanical parameters of the 100%RAP mixes could vary. In other words, each laboratory tested the 100% RAP specimens and established the X_{RAP} (T, test) at five temperatures with T = (70, 100, 140, 170 and 190 °C). In order to have consistency

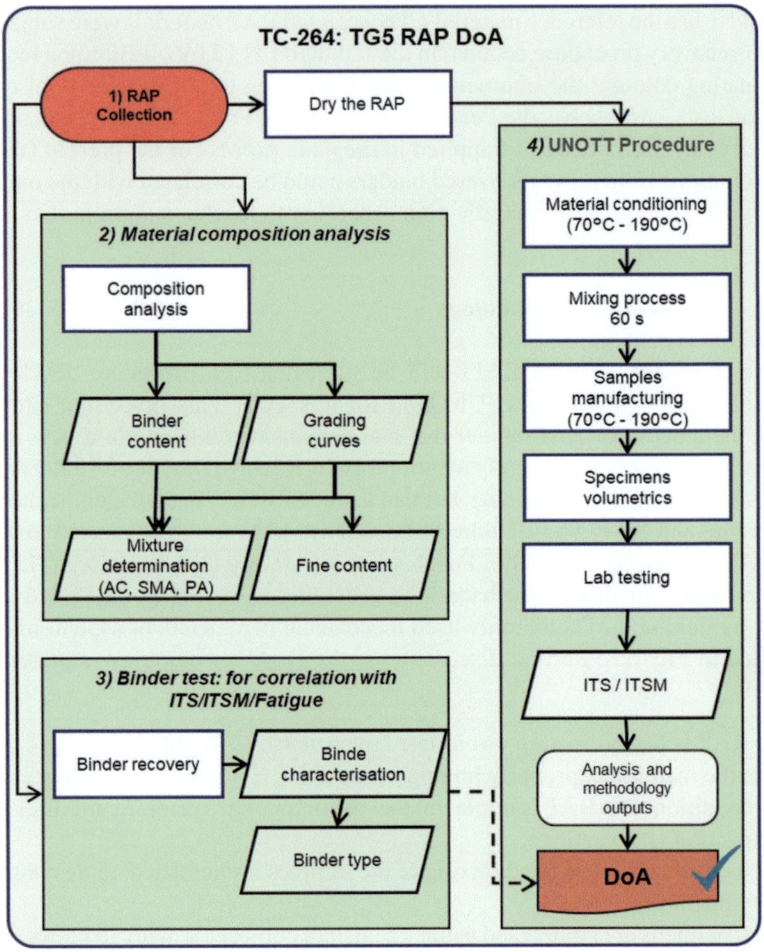

Fig. 5.8 TG5 procedure flowchart

between the laboratories, it was required that each lab could at least assess the Indirect Tensile Strength (ITS) of each specimen. Hence, receiving a series of X_{RAP} (T, ITS) was necessary. In recent studies conducted at the University of Nottingham [37], it has been shown that ITS of 100%RAP mixtures properties varies with the type of RAP and, most importantly, the selected range of temperatures. These same studies also allowed for estimating the relative DoA' via the calculation of the index in Eq. 5.1 with $X_{RAP} = ITS_{RAP}$ (T, ITS).

$$DoA'(\%) = 100 \times \frac{X_{RAP}(T, test)}{maxX_{RAP}} \tag{5.6}$$

where "X_{RAP}" is the ITS results of the RAP at a specific temperature "T", and the "max X_{RAP}" is the maximum ITS value observed among the temperatures evaluated.

TG5 proposed then to carry out a similar exercise amongst the participating laboratories by asking partners to assess at least the ITS at the relative temperature (T) of each manufactured 100%RAP asphalt mix. However, the framework is flexible enough that, based on the availability of material, time and equipment, each partner could have tested the same 100%RAP specimens with other laboratory tests.

5.5 Key Findings

5.5.1 'Relative Degree of Binder Activation' Definition

The terminologies used in this chapter correspond primarily to what was previously described in Sect. 2.4.1 and in Glossary. The term '*relative* degree of binder activation' (DoA') came from the discussions with the TG5 participants:

- It came from realizing that the results of DoA (the real Degree of Binder Activation) might depend on the selected test method.

DoA' could be defined as a relative level that expresses the "binder activation" as a function of the maximum achievable ITS value, assuming that when this is reached, there is 100% binder activation. Care should be taken since this seems to be not the case for the group of RAP that show a decrease in ITS values above a certain temperature, which shows that the "binder activation" is not the only parameter affecting this value. Further discussion and a possible distinction between the so-called DoA' and compactability should be addressed.

Also, it is still under discussion if ITS is the most appropriate test to access DoA'. In fact, when the same RAP is tested in the same lab, although evaluated through different mechanical tests (ITSM and ITS), the DoA' curves were not the same. The approach used on the TG5 interlaboratory effort looked at a simple test for RAP classification. However, some other tests might provide different answers.

5.5.2 Trends

There is a direct relationship between air voids and ITS (Fig. 5.9) and, consequently, the DoA'. The data seem scattered due to possible variance in the test methods applied to calculating the volumetrics.

In terms of the format of the curve ITS versus Temperature, three different trends were observed in the data from the 42 RAPs analyzed, as shown in Fig. 5.10a. The air void trend is inversely proportional to the ITS trend, as observed in Fig. 5.10b.

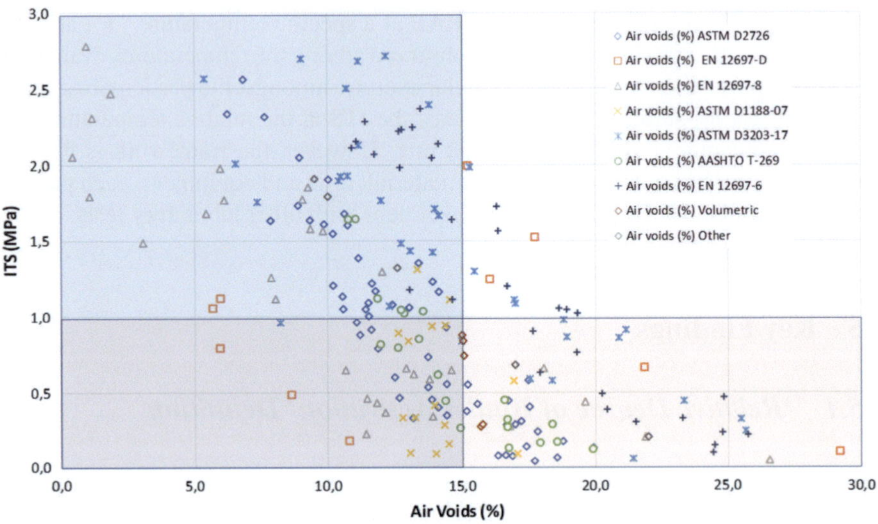

Fig. 5.9 ITS versus air voids

Analyses were performed to investigate if the 42 RAPs could be separated into clusters that could explain the different trends. The first attempt was made in terms of the laboratory and consequently, the test methods used and the type of asphalt binder, and it was observed that the ITS vs. Temperature trend is not directly related to the laboratory/test methods employed (Fig. 5.11a), neither the type of asphalt binder (Fig. 5.11b).

The second attempt was made with respect to the compaction method. Figure 5.12 below shows 6 RAPs compacted in the same laboratory through Marshall (50 blows per side) and Superpave gyratory compaction (SGC) (30 gyrations). Although the ITS results are different, the trend observed is the same, despite the compaction energy.

Some possibilities on why these different trends appear on the data are:

- 'over aging' of the RAP asphalt binder after a certain temperature causes the loss of bitumen's cohesive/tacking properties.
- A higher portion of the RAP bitumen becomes "extremely soft/almost liquid" in a way that causes modification on the internal structure of the mix.
- RAP clusters start to break after a certain temperature (dependent on the RAP), leading to a new skeleton rearrangement.

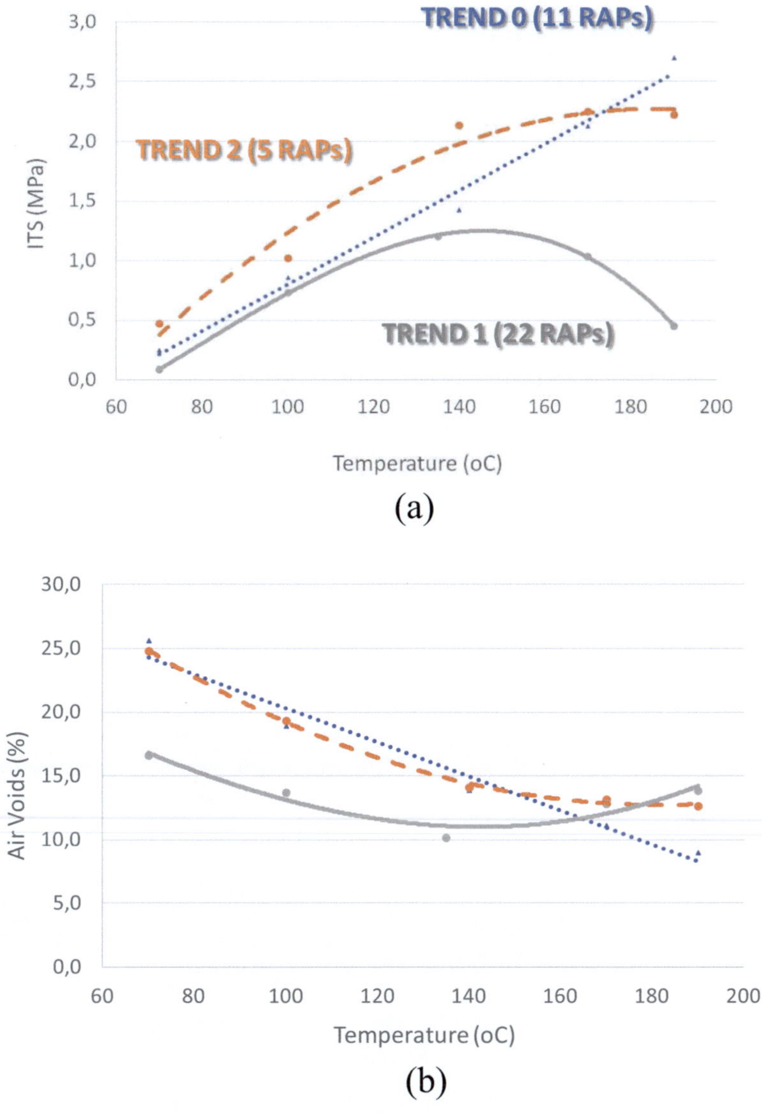

Fig. 5.10 Different trends observed for **a** ITS, and **b** air voids versus temperature (average results presented)

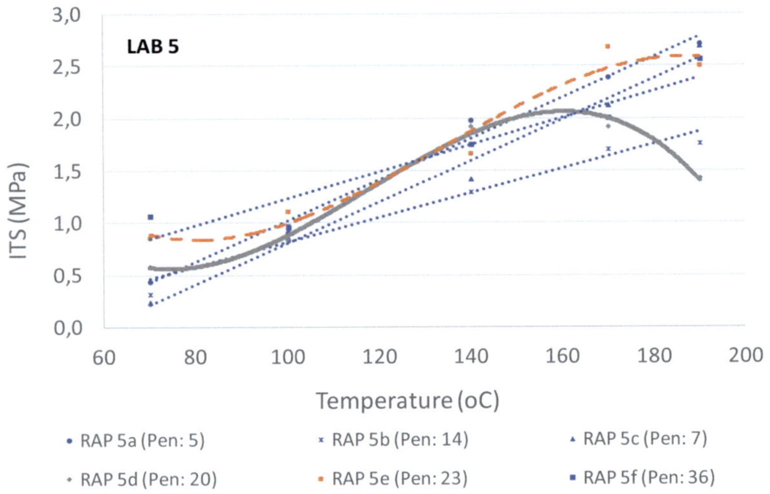

(a)

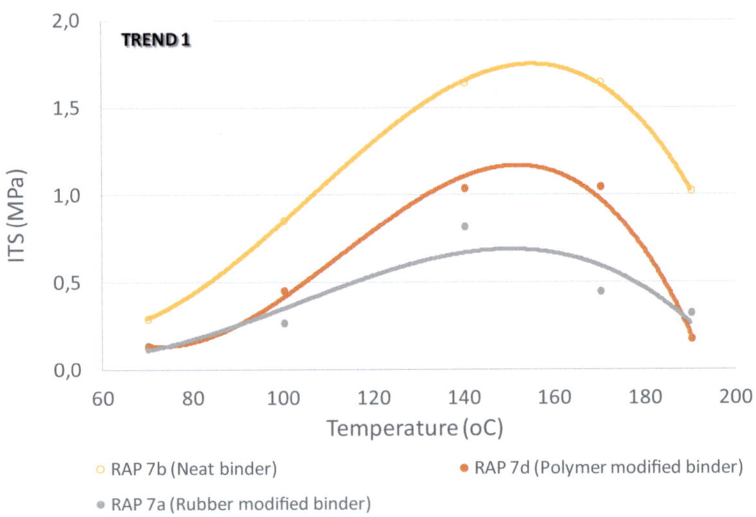

(b)

Fig. 5.11 **a** Different trends observed for RAPs prepared in the same laboratory (and same test methods), and **b** Same trend observed for different types of RAP asphalt binder

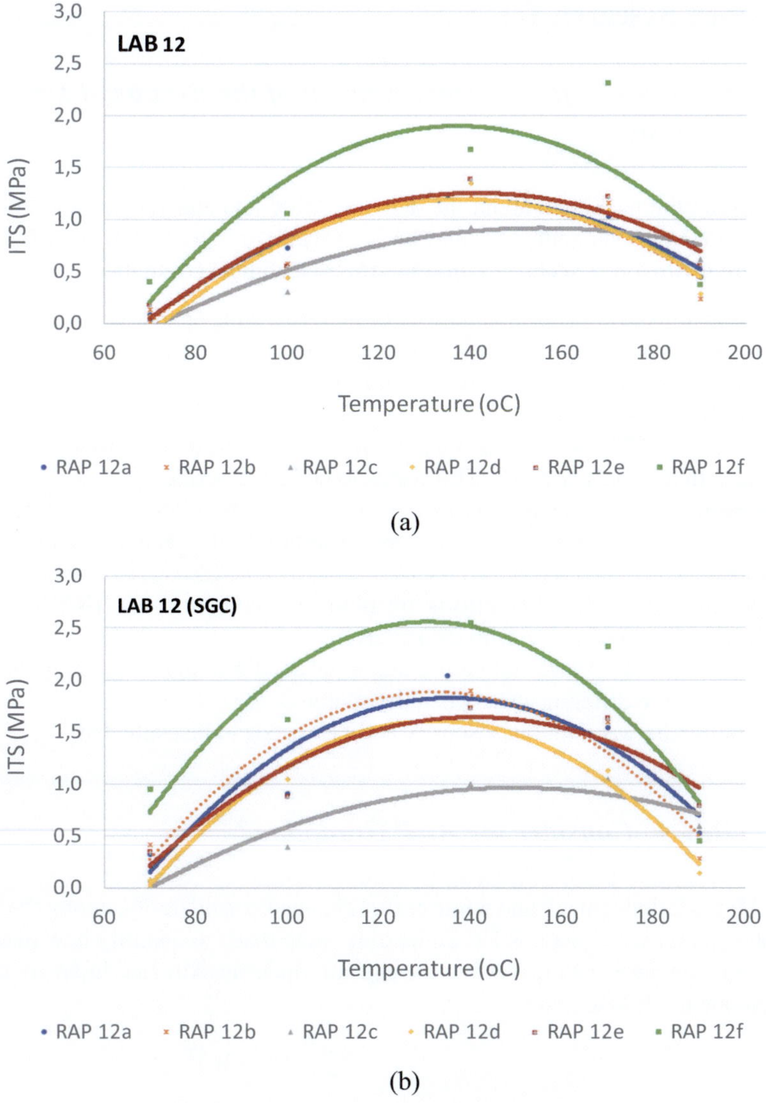

Fig. 5.12 **a** Marshall compaction (50 blows per side), and **b** SGC (30 gyrations)

5.6 Open Research Topics

5.6.1 *Suggestions for the Improvement of the Results of Data Analysis*

Much discussion happened about the data presented and analyzed. Some inputs in terms of how the data analysis could be improved are listed below, which should come along with a new technical group in future technical committee:

- Use probabilistic optimization method to analyze testing results, considering input data (such as temperature, gradation, and binder content…)
- Keeping consistent air void levels (that is, changing the compaction energy) for ITS at each temperature.
- The use of 140 °C is observed as a significant temperature in terms of substantial changes in air voids and ITS. This needs to be further explored.
- The evolution of the characteristics of the RAP binder after conditioning needs to be explored as it would provide some insight into the 'ageing' of the asphalt binder.
- Organizing data gathering efforts by nominal maximum size (NMS) groups (coarse and fine) can help with data analysis.
- Images of the RAP material before and after conditioning/compacting can give insight into the differences between the sources.
- Correlation between binder viscosity and ITS needs to be evaluated.

5.6.2 *How to Consider the RAP Gradation?*

The RAP gradations (black and white curve) changed considerably among the RAPs evaluated, as shown in Sect. 5.4.1. Some trials were made to quantify how much the RAP gradation curve changes when the asphalt binder is extracted in an attempt to differentiate the RAPs better.

$$\mathit{Gradation\ Index}\ 1 = \frac{\sum_{n=1}^{i} (B - W)^2}{i} \tag{5.7}$$

$$\mathit{Gradation\ Index}\ 2 = \frac{\sum_{n=1}^{i} [(B - W)^2]/\#}{i} \tag{5.8}$$

where i is the number of sieves, B is the percent passing on each sieve in the black curve, W is the percent passing on each sieve in the white curve, and # is the sieve mesh (in mm).

Figure 5.13a presents the results of gradation index 1, and Fig. 5.13b the results of gradation index 2. The information on the black and white curves was unavailable to all the RAPs. The different colours mean different types of asphalt as declared by each laboratory: blue—unknown, orange—neat binder, green—polymer modified binder, and yellow—rubber asphalt.

No clear trend was observed between the gradation indexes and the trends illustrated in Fig. 5.10. The change in gradation curves is undoubtedly associated with the activation of the asphalt binder, probably along with other factors (amount of binder, age of the RAP, among other factors). The gradation index may be very important

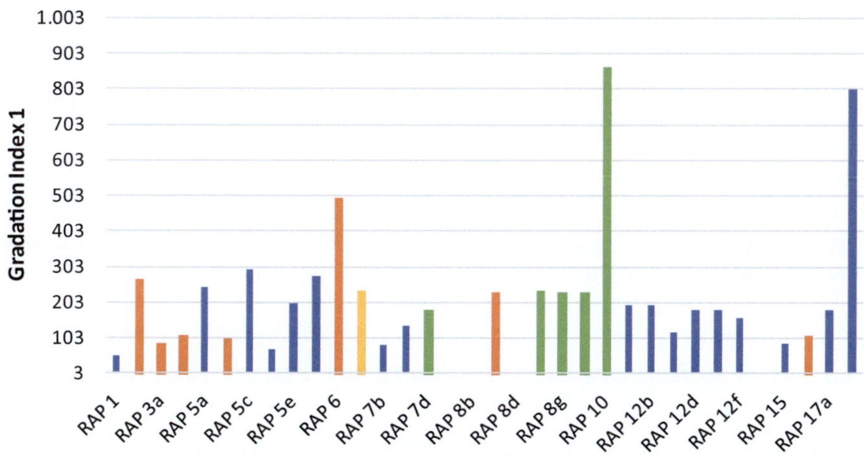

(a)

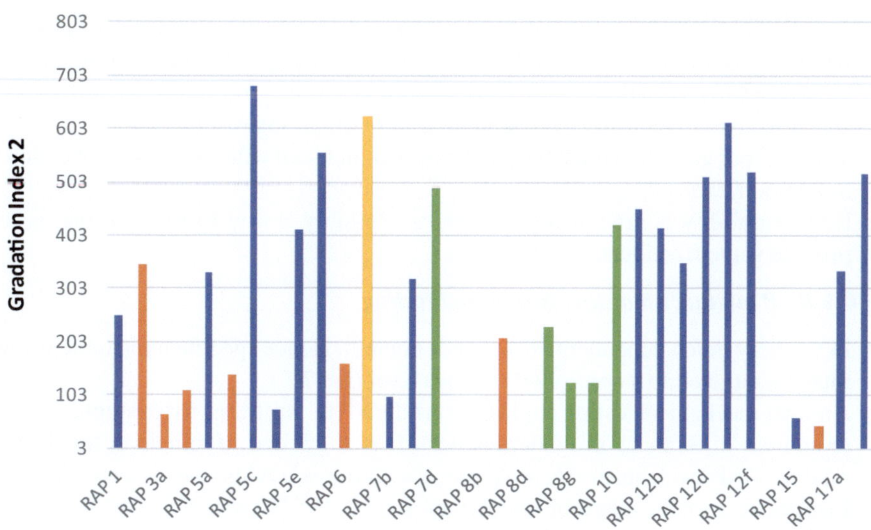

(b)

Fig. 5.13 **a** Gradation index 1, and **b** Gradation index 2

for RAP classification, and a massive space exists for investigation. Some of the possibilities discussed are:

- Separate coarse and fine RAP before the DoA' protocol evaluation.
- Associate the gradation index to just part of the sieves.
- Try some common and simple gradation quantifiers: the fineness modulus in Portland cement concrete design, the maximum density line (MDL) on the 0.45 power chart, and the coefficient of curvature or coefficient of uniformity from the Unified Soil Classification System (USCS) method (ASTM D2487).
- Define certain thresholds around the 4 to 6 mm sieve range, which might be able to split the RAPs into groups (open-, gap-, dense-graded).
- Check whether the two gradations (black and white) are somehow parallel. If the two curves are parallel, such as those in Fig. 5.7c, they will behave as similar mixture types (going from 0 to 100 DoA'). If they are different, such as the curves in Fig. 5.7c, this means that as DoA' increases, it changes from a more continuous type of mix to a more gap graded. The slope between the passing in the sand region between 1 and 2 mm can indicate mixture type change.

5.6.3 How to Improve the Interlaboratory Testing Program

Some points along the interlaboratory procedure can be optimized to facilitate data analysis:

RAP Material Selection

- Allowing laboratories to select their own RAP was practical and appreciated. However, this variability needs even more detailed procedures in 100%RAP specimen manufacturing and testing to make comparing results easier.
- The effect of gradation must be quantified and included with DoA' as a correction/ reduction factor.
- RAP variability is crucial for the results. Possibly a way to rule out this is to provide some fractionation guidelines.

100% RAP mixtures manufacturing and testing

- ITS is very much dependent on air voids content. Hence specimen manufacturing must be targeted to a controlled air void level (i.e. 7%)
- Another mechanical testing could be considered (i.e. stiffness, for example, seems to address differences in binders better). It is suggested to remain at the 100% mixture RAP level, but also Fine Aggregate Mixtures (FAMs) could be considered.
- Investigate another range of temperatures, 120–155 °C possibly, controlling air voids especially. Below 140 °C, the trend is similar to all RAPs.
- Evaluate the performance of recycling agents in a 100% RA-containing mixture and the possibility of enhancing the activation level.

Data Analysis

– The correlation between properties of 100%RAP mixtures, FAMs, mortar, mastics and binders should be better investigated. Probably machine learning can be of help.

Acknowledgements The authors are grateful for the support provided by following collaborators:

- Ahmed Abed (University of Nottingham, UK)
- Luciano Specht (Federal University of Santa Maria, Brazil)
- Leda Christiane Lucena (Federal University of Campina Grande, Brazil)
- Chibuike Ogbo (University of New Hampshire, USA)
- Kranthi K Kuna (Indian Institute of Technology Kharagpur, Kharagpur, India)
- Loretta Venturini (Iterchimica S.r.l., Italy)
- Gustavo Menegusso Pires (University of Nottingham, UK and Dynatest Latam, Brazil)
- Gaspare Giancontieri (University of Palermo, Italy)
- Amy Epps-Martin (Texas Transportation Institute/Texas A&M University, USA)
- Edith Arambula (Texas Transportation Institute/Texas A&M University, USA)
- Timothy Sobieski (Texas Transportation Institute/Texas A&M University, USA)
- Luis Miguel Gutiérrez Klinsky (CCR NovaDutra, Brazil)
- Marco Salles Chamouton (CCR NovaDutra, Brazil)
- Ramon Botella Nieto (Universitat Politècnica de Catalunya, Spain)

References

1. Kandhal, P.S., Mallick, R.B.: Pavement Recycling Guidelines for State and Local Governments (FHWA-SA-97) (1997). Last accessed from Washington DC, United States
2. Copeland, A.: Reclaimed Asphalt Pavement in Asphalt Mixtures: State of the Practice (FHWA-HRT-11–021) (2011)
3. Abed, A.: Mix Design, Mechanical Analysis and Performance Modelling of Warm Mix Asphalt. (Doctor of Philosophy), The University of Nottingham (2019)
4. Zaumanis, M., Mallick, R.B., Frank, R.: 100% recycled hot mix asphalt: a review and analysis. Resour. Conserv. Recycl.. Conserv. Recycl. **92**, 230–245 (2014). https://doi.org/10.1016/j.resconrec.2014.07.007
5. Lee, J., Edil, T., Tinjum, J., Benson, C.: Quantitative assessment of environmental and economic benefits of recycled materials in highway construction. Transp. Res. Rec. J. Transp. Res. Board **2158**, 138–142 (2010). https://doi.org/10.3141/2158-17
6. Zaumanis, M., Mallick, R.B., Frank, R.: 100% Hot mix asphalt recycling: challenges and benefits. Transp. Res. Procedia **14**, 3493–3502 (2016). https://doi.org/10.1016/j.trpro.2016.05.315
7. Baghaee Moghaddam, T., Baaj, H.: The use of rejuvenating agents in production of recycled hot mix asphalt: a systematic review. Constr. Build. Mater. **114**, 805–816 (2016). https://doi.org/10.1016/j.conbuildmat.2016.04.015
8. Lo Presti, D., del Barco, J., Carrión, A., Airey, G., Hajj, E.: Towards 100% recycling of reclaimed asphalt in road surface courses: binder design methodology and case studies. J. Clean. Prod. **131**, 43–51 (2016). https://doi.org/10.1016/j.jclepro.2016.05.093
9. Mogawer, W.S., Austerman, A., Roque, R., Underwood, S., Mohammad, L., Zou, J.: Ageing and rejuvenators: evaluating their impact on high RAP mixtures fatigue cracking characteristics using advanced mechanistic models and testing methods. Road Mater. Pavement Des. **16**(sup2), 1–28 (2015). https://doi.org/10.1080/14680629.2015.1076996

10. Willis, R., Turner, P., Julian, G., Taylor, A.J., Tran, N., Padula, F.D.G.: Effects of Changing Virgin Binder Grade and Content on RAP Mixture Properties, pp. 12–03 (2012)
11. McDaniel, R.S., Soleymani, H., Anderson, R.M., Turner, P., Peterson, R.: Recommended Use of Reclaimed Asphalt Pavement in the Superpave Mix Design Method (Project D9–12) (2000)
12. Mensching, D.J., Daniel, J.S., Bennert, T., Medeiros, M.S., Elwardany, M.D., Mogawer, W., Alavi, M.Z.: Low-temperature properties of plant-produced RAP mixtures in the Northeast. Road Mater. Pavement Des. **15**(sup1), 1–27 (2014). https://doi.org/10.1080/14680629.2014.926617
13. West, R.C.: Best Practices for RAP And RAS Management (QIP 129)
14. West, R.C., Copeland, A.: High RAP Asphalt Pavements: Japan Practice—Lessons Learned (IS 139) (2015)
15. McDaniel, R., Anderson, R.M.: Recommended Use of Reclaimed Asphalt Pavement in the Superpave Mix Design Method: Technician's Manual (REPORT 452) (2001)
16. Mastrofini, D., Scarsella, M.: The application of rheology to the evaluation of bitumen ageing. Fuel (1999)
17. Thom, N.: Principles of Pavement Engineering (Second edition ed.). United Kingdom: ICE Publishing (2014)
18. del Barco, J., Carrión, A., Lo, P.D., Airey, G.D.: Binder design of high RAP content hot and warm asphalt mixture wearing courses. Road Mater. Pavement Des. **16**(sup1), 460–474 (2015). https://doi.org/10.1080/14680629.2015.1029707
19. West, R.C., Tran, N.H., Taylor, A.J., Willis, R.J.: Comparison of Laboratory Cracking Test Results with Field Performance of Moderate and High RAP Content Surface Mixtures on the NCAT Test Track. Dordrecht (2016)
20. Hansen, K.R., Copeland A.: Annual asphalt pavement industry survey on recycled materials and warm-mix asphalt usage: 2009–2013. In: 4th Annual Survey (IS 138). National Asphalt Pavement Association, Lanham, Maryland (2014)
21. Dinis-Almeida, M., Castro-Gomes, J., de Lurdes Antunes, M., Vieira, L.: Mix design and performance of warm-mix recycled asphalt. Proc. Inst. Civ. Eng. Constr. Mater. **167**(4), 173–181 (2014)
22. Al-Qadi, I.L., Carpenter, S.H., Roberts, G., Ozer, H., Aurangzeb, Q., Elseifi, M., Trepanier, J.: Determination of Usable Residual Asphalt Binder in RAP (ICT-09–031) (2009). Last accessed from Illinois, United States
23. McDaniel, R.S., Shah, A., Huber, G.A., Copeland, A.: Effects of reclaimed asphalt pavement content and virgin binder grade on properties of plant produced mixtures. Road Mater. Pavement Des. **13**(sup1), 161–182 (2012). https://doi.org/10.1080/14680629.2012.657066
24. Al-Qadi, I.L., Elseifi, M., Carpenter, S.H.: Reclaimed Asphalt Pavement—A Literature Review (FHWA-ICT-07–001) (2007)
25. Lo Presti, D., Vasconcelos, K., Orešković, M., Pires, G.M., Bressi, S.: On the degree of binder activity of reclaimed asphalt and degree of blending with recycling agents. Road Mater. Pavement Des. **21**(8), 2071–2090 (2019). https://doi.org/10.1080/14680629.2019.1607537
26. Zofka, A., Marasteanu, M.O., Clyne, T.R., Li, X., Hoffmann, O.: Development of Simple Asphalt Test for Determination of Rap Blending Charts (2004–44) (2004)
27. Shirodkar, P., Mehta, Y., Nolan, A., Sonpal, K., Norton, A., Tomlinson, C., Sauber, R.: A study to determine the degree of partial blending of reclaimed asphalt pavement (RAP) binder for high RAP hot mix asphalt. Constr. Build. Mater. **25**(1), 150–155 (2011). https://doi.org/10.1016/j.conbuildmat.2010.06.045
28. Abed, A., Thom, N., Lo Presti, D.: Design considerations of high RAP-content asphalt produced at reduced temperatures. Mater. Struct. **51**(4) (2018). https://doi.org/10.1617/s11527-018-1220-1
29. Orešković, M., Pires, G.M., Bressi, S., Vasconcelos, K., Lo Presti, D.: Quantitative assessment of the parameters linked to the blending between reclaimed asphalt binder and recycling agent: a literature review. Constr. Build. Mater. **234**, 117323 (2020). https://doi.org/10.1016/j.conbuildmat.2019.117323
30. Bonaquist, R.: New Approach for the Design of High RAP HMA (2005)

31. Zaumanis, M., Mallick, R.B.: Finite Element Modeling of Rejuvenator Diffusion in RAP Binder Film—Simulation of Plant Mixing Process, pp. 407–419 (2013). https://doi.org/10.1007/978-94-007-6878-9_30
32. Coffey, S., DuBois, E., Mehta, Y., Nolan, A., Purdy, C.: Determining the impact of degree of blending and quality of reclaimed asphalt pavement on predicted pavement performance using pavement ME design. Constr. Build. Mater. **48**, 473–478 (2013). https://doi.org/10.1016/j.conbuildmat.2013.06.012
33. Navaro, J., Bruneau, D., Drouadaine, I., Colin, J., Dony, A., Cournet, J.: Observation and evaluation of the degree of blending of reclaimed asphalt concretes using microscopy image analysis. Constr. Build. Mater. **37**, 135–143 (2012). https://doi.org/10.1016/j.conbuildmat.2012.07.048
34. Bowers, B.F., Moore, J., Huang, B., Shu, X.: Blending efficiency of reclaimed asphalt pavement: an approach utilizing rheological properties and molecular weight distributions. Fuel **135**, 63–68 (2014). https://doi.org/10.1016/j.fuel.2014.05.059
35. Margaritis, Alexandros, et al.: On the applicability of ATR-FTIR microscopy to evaluate the blending between neat bitumen and bituminous coating of reclaimed asphalt. Coatings **9**(4), 240 (2019)
36. Hassan, N.A., Khan, R., Raaberg, J., Presti, D.L.: Effect of mixing time on reclaimed asphalt mixtures: an investigation by means of imaging techniques. Constr. Build. Mater. **99**, 54–61 (2015)
37. Pires, G.M., Lo Presti, D., Airey, G.D.: A practical approach to estimate the degree of binder activity of reclaimed asphalt materials. Road Mater. Pavement Des. **22**(5), 1093–1116 (2021). https://doi.org/10.1080/14680629.2019.1663244
38. Welllner, F., Werkmeister, S., Falla, G.C., Millow, R., Blasl, A., Di Mino, G., Di Liberto, C.M., Noto, S., Lo Presti, D., Jimenez del Barco Carrión, A., Airey, G.: No D2.1 High-content RA warm asphalt mixture design. ALLBACK2PAVE: toward a sustainable 100% recycling of reclaimed asphalt in road pavements (CEDR Call 2012: Recycling: Road construction in a post-fossil fuel society). Brussels (2015)
39. Campher, L.: Herwonne Asfalt: Hoe effektief is verouderde bitumen in Bitumen Stabliliseerde Materiale? University of Stellenbosch, Stellenbosch (2012)
40. Tebaldi, G., Dave, E., Cannone Falchetto, A., Hugener, M., Perraton, D., Grilli, A., Lo Presti, D., Pasetto, M., Loizos, A., Jenkins, K., et al.: Recommendation of RILEM TC237-SIB on cohesion test of recycled asphalt. Mater. Struct. Springer Netherlands, **51**(5) ,117 (2018). https://doi.org/10.1617/s11527-018-1238-4

Chapter 6
Summary, Conclusions and Future Needs

Gabriele Tebaldi and Eshan V. Dave

6.1 Conclusions

Detailed accounts of specific findings and conclusions for all five domains of activities undertaken in TC 264-RAP are presented at the end of Chaps. 2 through 5. Key conclusions drawn from the TC activities with respect to asphalt pavement recycling are:

- Through interlaboratory characterization using unique materials, it was found that: (1) conventional asphalt binder characterization methods as well as rheological characterization using measurements from the dynamic shear rheometer showed that the properties of the binder blends made with rejuvenating agents, recovered binder from RA and virgin binder can have response similar to those of the original virgin binder; and, (2) the ageing effect on the binder blend of virgin binder, recovered binder from RA and rejuvenating agent is comparable to that experienced by the reference virgin binder with similar trends for an increasing degree of ageing. Interlaboratory testing efforts also demonstrated that when different recovered asphalt binders from RAs (different sources) are used for preparing binder blends, different trends in the empirical and rheological response are observed that depend on the RA source and the rejuvenating action of rejuvenators.
- There are insufficient life cycle assessment (LCA) studies that account for different types of asphalt recycling approaches. It is supported by many studies that reclaimed asphalt materials lead to less environmental impacts over conventional hot mix asphalt.

G. Tebaldi (✉)
Department of Engineering and Architecture, University of Parma, Parma, Italy
e-mail: gabriele.tebaldi@unipr.it

E. V. Dave (✉)
Department of Civil and Environmental Engineering, University of New Hampshire, Durham, NH, US
e-mail: eshan.dave@unh.edu

G. Tebaldi and E. Dave (eds.), *Recycling Asphalt Pavement*, RILEM State-of-the-Art Reports 44, https://doi.org/10.1007/978-3-031-94231-0_6

6.2 Future Needs

The present conclusions, although based on a large interlaboratory activity, are centred on experimentation at the binder level and specifically on unmodified binders. To confirm further the potential of the proposed methodology and extended to research activity on the ageing behaviour of ARA, it is suggested to expand the investigation to more complex bituminous material phase such as mastic, mortar and mixture. Such a research effort could be addressed in a follow-up RILEM TC.

However, this is not an easy task as various methodological and technical difficulties exist in LCA for RAP. For the RAP LCA methodology, challenges regarding definition of RAP, choice of LCA scope, functional unit, system boundaries, end-of-life/end-of-waste, allocation methods, and recycling methods widely exist, so that the RAP LCA results include uncertainties and, sometimes, are impossible to compare with one another. It is difficult to use current literature to resolve these challenges. Hence, there is an urgent need to standardize RAP LCA by developing RAP LCA methodology and establishing universal PCR, EPD, and LCI. There are also various technical difficulties for RAP to be solved, in order to advance knowledge in LCA for RAP. These technical difficulties may include: a) Issues such as hardening of bitumen and re-activation of bitumen etc. are not well understood; b) There is no universal technical approach to evaluate pavement durability with different percentages of RAP or other additives (e.g. rejuvenator). Hence, it remains a question regarding how to increase the percentages of RAP used in the mix to balance durability and environmental footprints.

In summary, LCA is an essential tool to evaluate the environmental footprints of RAP and can help to optimize recycling methods to reduce those impacts. However, various methodological and technical difficulties exist so that their environmental footprint cannot be accurately or even fairly assessed. Many of these difficulties are not easy to deal with, however, more research in RAP LCA is essential and urgently needed, in order to promote cleaner production in the asphalt industry. Furthermore, standardized RAP LCA methodology is urgently need.

Glossary

Absorbed RA binder The amount of aged binder that is absorbed by RA aggregate and it is not considered effective as binder film.

Activated RA binder The activity of a certain amount of black rock RA binder and absorbed RA binder due to the combined effect of time, temperature and recycling agents.

Active RA binder The amount of RA binder made active during processing of selected RA due to the effect of time and temperature. It consists of two components: liquid RA binder and Softer RA binder and it equals the available RA binder only when no recycling agent is included.

Available RA binder The amount of binder that can be considered available in the formulation of recycled asphalt mixtures. This is composed by the active RA binder, eventually complemented by the activated RA binder and the residual recycling agent.

Black rock RA binder The amount of aged binder which has become so stiff and brittle that is considered as part of the RA aggregate, showing no change in physical–chemical behaviour and properties of itself.

Degree of binder Activity (DoA) An index indicating the minimum percentage of active RA binder that a designer can consider for a selected RA and a selected asphalt manufacturing process. DOA is an intrinsic property of each RA.

Degree of binder Availability (DoA) An index indicating the maximum percentage of available RA binder that a designer can consider in new mix design formulations.

Degree of binder Blending (DoB) An index describing to what extent the aged RA binder contributes to the final properties of the asphalt mixture's binder blend composed of aged binder and recycling agent.

Full availability (100% active RA binder) The total amount of RA binder can be considered in the new mix design as part of the new binder blend.

Full blending The available RA binder and the recycling agent blends perfectly, hence the property of the final binder blend/mixture are directly correlated to available RA binder.

G. Tebaldi and E. Dave (eds.), *Recycling Asphalt Pavement*, RILEM State-of-the-Art Reports 44, https://doi.org/10.1007/978-3-031-94231-0

Inactive RA binder The amount of aged binder that can only be partially activated by the combined effect of time, temperature and a recycling agent.

Liquid RA binder The portion of aged binder that moves from the RA particle to other RA particles, or virgin aggregates.

Long-term activated binder (LTAb) Long term activity is due to an eventual diffusion of the recycling agent into the RA binder over time. It might result in an even bigger decrease of the unavailable RA binder.

Null availability/Black Rock (0% active RA binder) Assumes that RA behaves as "black aggregate" and no RA binder can be considered in the new mix design as part of the new binder blend.

Null blending Available RA binder doesn't have any effects on the properties of the final binder blend/mix.

Partial blending A more realistic hypothesis that lies between the full availability and black rock scenarios. Hence only a partial amount of RA binder can be considered in the new mix design as part of the new binder blend.

Partial blending A more realistic hypothesis that lies between the full blending and null blending scenarios.

Relative Degree of binder Activity (relative DoA) An index indicating the minimum percentage of active RA binder obtained with a selected testing procedure.

Recycling agents The family of additives that are added within the recycled asphalt mixture manufacturing process to act on the aged binders with two different purposes: (1) "Rejuvenators", restore the properties of the aged RA binder (i.e. neat binder, naphthenic oils, etc.) and (2) "Lubricants", facilitate the mixing production process by allowing lower manufacturing temperatures, hence higher RA content (i.e. warm mix technologies).

Residual recycling agent Amount of remaining recycling agent that will stay in the blend as a residual component and maintaining the original properties.

Short-term activated binder (STAb) Short term activity happens when the recycling agent gets in contact with aged binder at fixed processing conditions and it results in a decrease of the Black Rock and possibly Absorbed binder.

Softer RA binder The layer of aged binder that does not move from the RA particles but becomes softer and acts as a glue.

Total RA binder Amount of binder potentially available as a binding agent and composed of the RA binder and, when present, recycling agents.

Unavailable RA binder The amount of aged binder that cannot be considered available in a new formulation: This quantity is made of two components: Black Rock RA binder and Absorbed RA binder.

FSC
www.fsc.org
MIX
Papier aus verantwortungsvollen Quellen
Paper from responsible sources
FSC® C105338

If you have any concerns about our products, you can contact us on
ProductSafety@springernature.com

In case Publisher is established outside the EU, the EU authorized representative is:
Springer Nature Customer Service Center GmbH
Europaplatz 3, 69115 Heidelberg, Germany

Printed by Libri Plureos GmbH
in Hamburg, Germany